Deep Six:

the final dive ...

By R. Douglas Veer

Changed Lives Publishing, LLC

2022

This book is based on a true story
but some names and places have been changed
to maintain confidentiality.

First Printing: 2022
ISBN-13: 978-1-7323243-5-0
Library of Congress Control Number: 2022912180

Cover Design: K. Veer
Front Cover Photo Credits: Jordan, David. Wolfpack:
The U-Boat War 1939-1945. New York:
Barnes & Noble Books, 2002
Reprinted at http://www.usmm.org/position.html

* * * * * * *

Book Ordering: Amazon.com
Search: R. Douglas Veer
Booking Information, please contact:
www.KathyVeer.com or www.DougVeer.com

Changed Lives Publishing, LLC
Region of Augusta, Georgia

PREFACE

This book is not as much a biography of my life as it is an account of life aboard a submarine and the men who willingly commit their lives to such a calling.

When my wife asked me to write about my experiences as a submariner I hesitated, because I felt I didn't have much to say. To me it was so simple: I joined the U.S. Navy, I volunteered for submarine school and was accepted, and I happily and gratefully served aboard three of those marvelous, cigar-shaped floating machines of war, designed to keep the peace.

This work reveals some of my life's endeavors as a serviceman and brings out the stories of many others like myself, who served with the same crazy, dignified, devoted love. This kind of dedication is what's required of the men and women who serve, in an attempt to keep the peace not only for our country but for the rest of the world.

The following Bible passages found in Psalm 107:23 (AMP) and Isaiah 42:10 (NIV) say,

"Those who go down to the sea in ships, Who do business on great waters; They have seen the works of the Lord, And His wonders in the deep . . . Sing to the Lord a new song, his praise from the ends of the earth, you who go down to the sea, and all that is in it . . ."

I hope I have done that.

R. Douglas Veer

CHAPTER ONE

THE GREAT DREAMER

Who, at birth, can possibly know where the next 60, or 70, or more, years will take them? The Bible says we are given three score and ten; that's 70 years, and we are given extra years above that if we honor our fathers and our mothers. But during those years, who can say what we'll do to shape our paths in life or what will happen as a result of the decisions we make? In one sense, we are our future.

As a nine or ten-year-old boy, he would often lie in the grass beneath a shade tree on the north side of his home and stare into the sky above, wondering what was on "the other side", as if he were inside a great ball and couldn't see the outer portion of it.

But as he continued to daydream--considering that even if he could one day get "outside" to see beyond his restrictions--maybe he would find another layer beyond what he could see, and another beyond that, and another, and yet another. Perhaps there were balls inside of balls that went on forever, always unknown, yet always provoking him to wonder what was beyond.

Yes, that young man was me. I was a great dreamer and such were the dreamy thoughts I often had while relaxing under the coolness of that shade tree on the north side of my house; an open book lying beside me.

Leaning against that great tree, I could see across the pasture where there were more trees, and tall fuzzy cat-tails, and just past that were boysenberry bushes, and mulberry bushes, obscuring the view of a few cows, lazily munching grass on their side of the fence.

Unlike me, nothing in the world bothered them. They were content with the simple fodder in front of them, uninterested in what was beyond their noses. I, on the other hand, dreamed of far-off places I might go someday, far from the pasture, the cat-tails, and the berry bushes.

I often fell asleep with those dreams in my mind and the open book next to me that I hadn't finished reading. I imagined *the future* and the places it would take me. I wondered, "Would I ever get to go beyond this great blue ball? Would my life ever take me anywhere but here? Or would I just lie here, grow old, and die, never knowing what was outside of my tiny part of the world?"

ABOUT GROWING OLD

We all know it's a fact that some people live to be 100 years or more; people such as my mother who died only days before turned one hundred years young. Because even after giving birth to, and raising fourteen children, she did seem to be eternally young at heart.

But that is not the normal life span for us humans. Even if you consider infant death or accidents, many of us have not lived to enjoy the 70 years promised, according to the Bible. So the question for us becomes, "What do I have to look forward to as I peer into the long, dark, tunnel of my life as it stretches out before me? What is in the *beyond?"*

As I grew older, my world expanded past my small, *shade-tree speculations* of life. The world around me was more and more fascinating as I ventured into the woods and pastures, and it was much larger than I thought it to be.

As I wandered farther from home, I climbed the hills from which I could see into the valleys. I saw streams, rivers, and places in the distance that were once limited horizons of a child's life. And as I saw more, I wanted to discover more.

KEEPING IT ALL TOGETHER

Then, as the years passed and I became older, I was distracted from my boyhood thoughts by the reality of living life *in the now* and trying to find my place in it. The daily and hourly, mundane grind of living became my focus, which sadly, becomes a necessity for all of us.

During my teens and early twenties, there were several years of *footloose* living; I was wild and irresponsible and did whatever came along. But then, in my twenties, and thirties, I began facing those responsibilities, caring for myself, a wife, and a child, as they took first place in my life along with providing a living for them and covering the many expenses.

No longer were my thoughts captured, wondering *what is ahead*, or thinking, *if it feels good do it!* I realized I *had to* do it . . . like it or not.

Now, I could only think about *what is now* and what I needed, to keep life and family together. I was no longer daydreaming about the years ahead. My *big blue ball* was slowly forgotten; the *great, mysterious ball* of life faded from my sight, disappearing somewhere into my forgotten past.

We all have that road to walk. But some of us, like me, couldn't see where it was leading. I didn't have the time, or insight to even think about where that road led. At that point in life, I could only think about my family, myself, and, *"What do I do now?"*

WHAT THE FUTURE MAY HOLD

My future--the years that were waiting, silently, for me to catch up to--were hardly considered anymore. Yet, to be honest, every once in a while I would again wonder about that forgotten road and what it might still hold if I would, or could pursue it. But for now, I had to live my life for the sake of others!

Once in a while, time would pull me back to that place, as I returned to that shade tree to gaze at the horizon and dream dreams of the future . . . but even those dreams were fading as *real-life* overshadowed them.

As I struggled back and forth I could not completely let go of the idea that I was just like everyone else. I *did* have a life ahead of me, and a future to embrace. I *should* be able to know what that was going to be. And somehow I *could* go beyond the small pasture with cows and the mulberry bushes. Couldn't I?

Surely there was a future out there, and it was there for a reason. It would be a terrible loss if I allowed it to waste away, to be forgotten, never to be explored and enjoyed. Right?

But was it still possible? Could I still find those boyhood dreams? Could I still find the answers to my questions? Perhaps. But first, let me tell you about those earlier times before I became bound to marriage, children, and other adult responsibilities.

LIFE IS REAL

In my mid-teens, I joined the military and it afforded me the privilege of traveling to other countries where I discovered the oceans which not only stood between us but also joined our country to many others. I was a young man and my world was rapidly expanding beyond that tiny world I once thought I lived in.

Any way you look at it, life is tough. Rich or poor, black or white, boy or girl, young or old; it's tough. But, I knew I could go on. After all, "Life is what you make of it," right? I've often heard that said although I'm not sure I fully agree with it.

My friend and mentor, one-time Pastor, and pseudo-adopted father used to say something as close to what I agree with as anything I've heard. He would say, "life is real". It was simple, it was without judgment, and it was hard to argue with. Chewing on that would give me plenty to think about. And the older I've gotten the more I agree with him. Yes, "life is real" but I would add, "The real problem with life is, it goes by too fast."

It was only yesterday that I was a child, worrying about nothing, playing, and having fun. That of course is in addition to cleaning up the yard, taking care of rabbits and chickens, working my paper route, walking a half-mile to the store and back to get things for Momma to fix for Dad's lunch, and going to school. Life was a *bowl of cherries,* right? Yes, but only after all the chores were done and my school work was taken care of. And . . . it seemed there was never enough time left over just for me.

Yeah, I thought life was tough. But as I look back I must ask myself, "Was it?" What did I know? I was young and I had a lot to learn. I was just *bull-headed* and stubborn enough to be determined I would make it anyway.

A MILITARY FAMILY

My father was never in the military but my two older brothers, Bill and Frank, were in the Navy during World War Two. Next in line was my brother Karl who followed them by going into the Air Force. He was followed by my sister, Neva, who married a submarine sailor, and she was followed by Charles "Bud" who joined the Marines. After that, my brother Ernie joined the Air Force. I was next in line, becoming a Submarine sailor because of being so captivated by Neva's husband, "Dutch".

After me were my sisters Rosemary, Beatrice, Grace, and Charity, and my younger brothers, John, Sam, and Ronnie. They all served or married into a U.S. military family. Eventually, they all had children who also honorably served, going all across the face of the earth from sunrise to sunset: Germany, China, Japan, Korea, Viet Nam, Desert Storm, Iraq, Afghanistan, Syria, the Persian Gulf, and anywhere else on earth the weak needed to be defended and protected; my family was there.

Fourteen of us in my immediate family all served and were dedicated to defending the freedoms given to us by the sacrificed lives that went before us. Our soldiers fought for this wonderful country called America while sending the military to other countries that were too small or too weak to defend their own freedoms.

Oh, do I sound as if I'm on a *soapbox* for the U.S. Military? Yes, thank you. I am!

May God continue to bless the U.S.A. as we bless the many citizens across this nation and beyond.

THE PIN-SETTER

As a boy I was pretty wayward, I was hard-headed, and sometimes my temper was hot enough to singe the hair off a billy goat. I guess that's why I rebelled against my father and left home at fourteen years old. It wasn't a wise move, and I certainly wouldn't suggest it to anyone else. But thankfully I had just enough grit to make it work. And work, I did!

Every day after school, and on Saturdays, I worked in a meat market, sweeping floors, washing trays, scrubbing counter tops, cleaning the inside glass in the meat cases, picking up and cleaning up behind the butchers, and even learning the trade of meat cutting.

And when my day was done there, I worked until dark setting up pins in a bowling alley that folks had knocked down. (That was before the pin setters became automatic.) Through it all, I stayed in school until just before my sixteenth birthday. That's when I contracted acute hepatitis that put me in the hospital for four months.

I was seriously ill but Momma saw to it that I was taken care of. And the wonderful men who owned the G & M Meat Market where I worked took care of me, as well.

Every week Johnny Grant, one of the owners, brought a basket of fruit to my hospital room with my paycheck sticking out of it. It was always my full pay. Never a week was missed. He treated me like I had actually put in my hours and days on the job just as if I deserved it. It's amazing how much I learned from his respectable show of kindness.

THE TATTOO

How did I get this sickness? I had my own money from my job at the meat market, I had my own room in a rooming house, and I had the freedom to do dumb things. One of them was when I attended a carnival that came to town.

I rode all the rides and played all the games. And yes, I got sick to my stomach several times. Then I saw a man putting tattoos on people and I thought it was cool, so I decided to get one, too. I watched for a while and decided to have my nick-name, *Doug,* tattooed on my upper, left arm.

Yeah, I was cool alright, but not so cool as to know about dirty needles and the infection they could carry. So, along with the tattoo, I got hepatitis. That, of course, was at no extra charge.

I was in the hospital over November 18, which was my sixteenth birthday, and I was still there on Thanksgiving, Christmas, New Year's day, and Valentine's day which was Sunday, February 14.

When I was released, my oldest brother, Bill, and his wife Charity took me to live with them on their farm until I could fully recuperate. When I finally did get well I went back to both of my jobs at the meat market and the bowling alley. I worked through that summer and proved I had become completely healed and healthy from hepatitis, and that's when I decided to enlist in the Navy.

On November 22, 1954, just four days after my seventeenth birthday, I enlisted and was accepted into the United States Navy.

I looked forward to finding a new life, a new path, and a new road that would take me to new places I had only dreamed of as that young, starry-eyed boy laying peacefully in the grass beneath the trees; that safe and secure home of my youth.

This was just the beginning and I was going to discover that *life is real* just like Dad said. Real? Yes. Real hard, and I was about to experience some of it.

LIFE GETS LARGER

My father had to sign papers permitting me to enlist. I don't think it was the easiest thing for him to do. Now that I am grown, and a father myself, I would be very apprehensive about allowing my children to go into military life. I would, of course, be proud if they did, but for me to sign them over and give my permission, would be hard, I know.

My brothers before me had gone to war and came back safely, yet I'm sure each time it was a little bit tougher for Dad to see another one go off to battle. I'm sure as each one of his nine boys came to that age, it was more and more difficult for him to sign them over.

Maybe it was his own sense of dedication to our country that caused him to let us go because that evening--there at the dining room table--he did complete the forms giving the Navy Recruiter permission to draft me into the service.

From there, everything happened pretty fast. I was given some kind of a ticket to ride the train and late the next evening I was seated next to some stranger, heading for the Great Lakes Naval Training Center near Chicago, Illinois. You sure were right, Dad. *Life was becoming real.* Very, very real.

EVERYTHING CHANGES

A whole new world was looming ahead and I'm sure I had thoughts like, "Why have I done this?"

Yet next to those scary thoughts were imaginations about how life was going to change for me. No more lying in my bed in my rented room, reading comic books and eating candy bars until my own chosen bedtime. No more holding hands with Caroline Knodle in the afternoons after school as she walked with me to my job at the meat market. No more deciding where to go, when to go, and why to go, because now I was no longer free to make my own choices.

My life was heading for dramatic change and although it was very exciting I also had a certain amount of apprehension welling up inside. Did I say apprehension? It was probably more like a deep, gripping fear; fear of the unknown; fear of what that unknown would become for me.

I wondered, "How hard was it going to be to learn what I was there for? What would be expected of me? What would happen to me when I made mistakes? (because I knew I would make them). How would I be treated *then?* How would I get along with the other men? Would there be arguments that would lead to fights? Surely there would be! Wouldn't there?" I had no one to turn to for answers.

My mind was buzzing with all those thoughts. I was finally facing the fact that I was really in the United States Navy and I was *their G.I. property;* Government Issue. The *big* question that continued to bug me was, "Have I made a mistake?" Even if I had, it was too late now because, "Yeah, Dad, life is real."

CHAPTER TWO

THE JOURNEY BEGINS

After Dad signed the papers, the recruiter gave me a pass to get on a Chicago-bound train the next day; I boarded it sometime around nine or ten o'clock at night. About half way to Chicago I found there were several other young men on that train who had just enlisted in the Navy, just like I had. I felt they all looked older than me, but they probably weren't.

It was about two o'clock in the morning when we arrived at our destination. The train was met by someone in uniform who took us to the actual Great Lakes Naval Training Center. From there, we were taken to a place they called *the mess hall* and we were offered hotdogs and something to drink. I didn't consider it too appealing since it was meant to be either a very late supper or a very early breakfast and either way I wasn't in the mood for *hot dogs.* But I was somewhat hungry as were the others so we ate what they gave us.

Next, we were taken to a place to sleep and we knew that after only a short couple of hours we would be called to get up again. Life for us as we had known it was about to come to a screeching halt. That *old life* was going to go through some very drastic changes. But, little did we realize just how drastic they'd be, even though we were now getting a taste of *real* military life. In less than five hours, we were going to start finding out.

REVELRY

There was no bugle blowing like I've seen in the movies; just someone yelling for us to *roll out for muster on the grinder*.

My first thought was, "Where is the grinder, and what is it?" My second thought was, "When are we going to eat?" Answers were quickly unfolding.

We were given four minutes to get dressed and then get outside to the huge blacktopped area, (the size of a shopping center parking lot). That is what a *grinder* is. It's where they grind you to pieces and put you back together as a military man should be; one who will become worth something to our country. I didn't understand it then, but years later I came to appreciate the process.

That's where it all starts: with waking up in the morning. And it's an adjustment everyone has to make in every single basic training class.

When you first arrive, the drill instructors generally have to use a lot of yelling to get everyone out of the bunk. But little by little, after a few weeks, (and the fewer the better), all it takes is for someone to come into the barracks every morning and say, "Roll out. Time for muster!" and immediately almost everyone is up and out of his bunk and lined up *in formation* on the grinder. Or at least they better be.

Once assembled there, the grinder is where a lot of basic calisthenics take place. That's their way of *grinding* all the nonsense out of you that you came in with. Plus, it makes every man equal, and it begins the process of making you as physically fit as you can possibly be. It's an amazing process but somehow we all make it.

DOING THE BEST I COULD

On this terribly uncoordinated first morning, they managed to get us lined up, fashioned into rows (ranks), and told to stand at attention and be quiet. We automatically realized that it was an order and by all means, we had better obey. Learning to obey is part of the process, too.

I didn't know it then but a year or so later, when I was on board a submarine, all of this was going to be very important to me and to the entire crew. Working together, peacefully and efficiently, was going to be very important and we were about to be taught how to do that, and why.

It's strange how just a few hours before that first day in boot camp, I wouldn't have paid any attention to such commands from someone I didn't know. But suddenly I was doing the best I could to stand straight and be quiet, which is called, *being at attention.* As time passed I would find that *the best I could* do would be made better and I would be thankful for that, too.

We were told how we would march to the mess hall for breakfast and when we were finished eating to sit quietly until someone came to take us back to the *grinder* for further orders.

I had no idea what Navy food was going to be like, but actually, it tasted pretty good. That morning it was chipped beef and gravy on toast and I thought to myself, "Yep . . . I'm here. I'm in the Navy." I was thrilled and halfway scared at the same time, but I was here so I told myself, "Eat up. There's a long day ahead."

THE PROCESS

I'm sure most people have heard about what goes on during the twelve weeks of preparing a man or woman for the next two, or four, or twenty years of military life, (which ever length of time one chooses to serve). This recruit training period or *boot camp* as it's called trains men and women to put aside the civilian life they are so accustomed to, and become a one-hundred-percent military person. They are made ready to serve anywhere in the world they might be called. The free-wheeling *civilian* person and their, *I'll-do-my-own-thing* attitude, is washed out of them. It is washed out, worked out, trained out, drilled out, taught out, or even sometimes, (to a certain degree), beaten out, because it is vitally necessary that the old independent nature is changed so one becomes a person of obedience to all higher orders and authorities, knowing that *their way is the best way* for one and all.

Without that change, we would never have an efficient military force. The *know it alls* and the *smart alecs*, the *disobedient*, and the *trouble-makers* are taken to a place they had never been before and they are changed into becoming a part of a cohesive fighting force that is meant to protect and defend liberty for anyone, anywhere on earth.

In essence, they are molded into the greatest, most efficient, most protective army in the world. In my case, it was the U.S. Navy and I had four years ahead of me to find the many answers I was looking for.

But first, a little sleep. It had been a long, confusing, complicated day and I was dead tired. We were taken to the barracks and shown some bunks and I realized again that I was on a military base, and in the United States Navy. Now it starts.

THE DISCIPLINE BEGINS

Once we were back on the *grinder,* I stood there thinking about the rented room where I'd lived the past couple of years and the many freedoms I had. There, after my work day was done, I could lay back in my bed, eat candy and read comics until I was tired enough to fall asleep. Then, in the morning, my alarm would wake me and I was off to school. When the school day was over it was back to the comfortable jobs that busied me until the end of my day. It was a pretty nice life for a teenage boy.

Then I thought back even farther to the room my brother and I shared at home where we lived happily with Mom and Dad and my brothers and sisters. My thoughts were a whirlwind of memories; memories from not that long ago, because I was just barely seventeen years old. It was only a few days ago that I was still sixteen and waiting for my seventeenth birthday. I was barely a child in many ways but here I was, ready to be a man, I hoped. I stood there daydreaming about the fact that my world had changed.

"SHUT UP AND STRAIGHTEN UP!"

From out of nowhere a booming voice said, *"You! Recruit! Straighten up and get your head out of the clouds and get with the program. Now!!"*

I was immediately pulled back to the reality of where I was and what I was supposed to be doing. When I looked at the Chief who was our drill instructor he was looking straight at me. Needless to say, my daydreams vanished into thin air.

I'm sure you've seen movie actors like Gomer Pyle playing this same scene for you, showing the instructor yelling at the men, telling them how to stand, how to be quiet, how they had to obey whatever they were told, and exactly how to do so without asking questions.

Just like in those movies, I soon learned how everything was going to be done, whether I liked it or not, so I'll spare you the details. Just know this: my training to become a military person had definitely started and it was way too late to change my mind.

THE "MAKE-OVER"

We were marched, (if you can call it that) to another building where we were stripped to our skin and sent to the showers. Once we were dry we stood in lines and walked past long tables where we were given underwear, socks, pants, shirts, and shoes. And before we were allowed to put them on, we were weighed and given a thorough physical examination from head to toe.

Next, we were given vaccinations, (for what, I'm not sure), and then we were allowed to get dressed and were sent outside in those new Navy *dungarees* to quietly stand in rows. When the last man came out we were once more commanded by our instructor how to stand in *rank-and-file* order, and *"BE QUIET!"* (And believe me, he meant business.) This was no *Gomer-Pyle-type* experience. This hardened old Navy Chief meant every word he said, and we knew it.

THE HAIRCUT

Doing the best we could, we *marched* to another building where we were taken into a long hall and told to stand there until we were called. One by one, and in groups of two or three, we were taken in to sit in a barber's chair. There, in what I will guess to be between fifteen and twenty-five seconds, we were given a *buzz* haircut. We were left with probably less than one-eighth of an inch of hair on our heads. Whether we had come in with hair to our shoulders or had a standard crew cut, we all got our hair *buzzed* nearly to the scalp.

Why would they do that? Number one, any diseases of the scalp and skin were exposed and treated immediately rather than passed on to others. Second, we were all made to become *alike* and made to realize who was *in charge* of our lives and every decision pertaining to such. In essence, it's a way of *breaking* each and every *unruly* aspect of our personality without actually destroying the talents, skills, and qualities we each possessed. We were being *honed* into strong, obedient, fighting military people.

I'm sure, more than anything, we each wanted to be part of that great United States military team.

Were there some who would refuse to give in to this one-for-all-and-all-for-one idea, to be part of something bigger than oneself? Yes. For those who were going to be nothing but trouble, they were generally given an *undesirable* discharge and sent home, never to be allowed to re-enlist, unless under very special conditions, and would have no military benefits to take with them.

Back then I didn't know what a great loss it would be not having V.A. medical benefits. But now, since I've enjoyed this wonderful privilege I realize how much I would have missed and I'm very grateful for those benefits today. Looking forward from then to where I am now--a senior citizen--the government has probably spent a million dollars or more on me just taking care of my health. I'm sure glad I stayed.

CHANGES

Boot camp consisted of training with weapons, receiving correction, and physically building and strengthening our bodies into active, aware, disciplined, healthy machines. In a very few short weeks, each and every *machine* was groomed, honed, and kept under control by someone sharing years of experience with us.

For many of us, the transition from a *snotty-nosed street kid* to a *standing at ready* military person was tremendous. It's not that I was really that bad but as with anyone else, we all had to be formed into a *one-unit team* of whichever branch of the military we became part of. And it was good.

OUT ON THE TOWN

About halfway through finishing the basic training we were allowed to take a night off and go into town. This time off is called *going on liberty*.

When we returned from *liberty,* we were once again measured and given more clothing. This time we received a dress uniform and shoes, which would complete our outfitting; it's what made us stand out as genuine U.S. Navy sailors.

These weren't just any old pants, tops, and shoes that we received. Along with these items came a special *pride* that any man or woman in the service would understand; they'd know exactly what I mean. Whether in the Army, Navy, Air Force, Coast Guard, National Guard, or Marines, all of us come to that very special place where we realize we are part of something grand; it leaves us humbled and proud at the same time. May God bless our troops at home and abroad and may He always keep them in the hollow of His hand.

MENTAL ABILITIES

Our one night of *liberty* was over all too soon and then we were back to our training schedule. The following week we were tested mentally as well as physically and we were given aptitude tests to discover which category we would best fit into, whether it be cook, mechanic, engineer, pilot, or bookkeeper. After a series of tests, some of us were offered schools according to our basic skills.

I don't know how it happened but I had an extremely high I.Q. which qualified me for some of the finest schools available. They offered me Officer Candidate School, at Newport, Rhode Island (also known as O.C.S.) and Naval Air Cadet School (NAVCAD) located in Norman, Oklahoma, or a school of my choice.

I was confused because at that time, I had absolutely no idea of the magnitude of these schools and what they could mean to me later. Graduating from either of those two schools would qualify me to become a junior grade Navy Officer, also called an Ensign. After that, I could be promoted to Lieutenant Junior Grade, Full Lieutenant, and so on up the line.

I guess at the time, I wasn't able to see the long road ahead; only the immediate outcome which was temporal. Besides that, all I could imagine was being in submarines, just like my brother-in-law, Dutch.

TIME FOR A DECISION

Suddenly, it was over. Boot camp was finished and we were given our various assignments. I often look back and wish I would have had someone to mentor me in making the many decisions a young man is confronted with, but I didn't.

So, rather than take one of the two tremendous school opportunities offered to me, my youth and my ignorance guided me, and I chose commissary school in San Diego, California.

Now, years later, I have regretted that decision. If I had it to do over I would have gone to NAVCAD, graduating as an officer and a pilot. Or I could have accepted OCS to become an officer, and then went on to volunteer for submarine duty. I could have continued with a twenty-year career in the military. But I didn't.

STAYING POSITIVE

I know I can't make it up to myself for making that mistake, so today I try to be a mentor to young people who have no one to help them and who are searching for advice about those life-changing decisions.

We, who have *been there and done that* so to speak, can and should--if asked--share our life experiences with those who come behind us. Especially if we can see they are struggling to find direction in their lives. Guiding a young person toward safe and productive outcomes and watching them grow can be one of the most rewarding experiences of our lives. What else is our experience worth if we don't pass it on?

CHAPTER THREE

WHY DID I DO IT?

Since I had worked in a meat market before my enlistment, I gravitated toward the school that was associated with food, which was the commissary school there in San Diego. It was also the training I needed to become a cook as Dutch had been.

When my training was over, I volunteered for submarine duty and was accepted. Soon, I was on my way to submarine school in New London, Connecticut, (instead of walking that higher road that would have gone way beyond cooking on a submarine).

But, I did what I had in my heart at the time and I've never been ashamed of it, only slightly saddened that through ignorance, I passed up such a golden opportunity.

"DUTCH"

As I said, my older sister, Neva, was married to a submarine sailor, Edward *Dutch* Gaulrapp, whose submarine was in port in Pearl Harbor, tied at the dock, on Sunday, December 7, 1941, when it was bombed by Japanese Aircraft. Dutch was baking bread when they got hit. And from there he was swept into World War Two as President Franklin D. Roosevelt declared war on Japan.

On that terrible day, 2,335 military personnel were killed. The count is 2008 U.S. sailors, 218 Army soldiers, and 109 Marines. Most of them were sailors simply because it was the ships that the Japanese planes were bombing, hoping to cripple our military strength. How wrong they were.

I had great respect for Dutch and because of him I felt drawn into submarine service and also followed him by deciding to be a cook. So, I passed all my tests and was accepted into submarine school, and then later graduated from commissary school. I had begun to feel like I was on my way to something very special.

SCHOOL-BOY AGAIN

Being part of this fine institution was beginning to make sense to me, and yet, it was also quite intimidating. Many had gone before me and I wasn't sure if my feet were big enough to fill the shoes of those whom I admired and respected so much. But I was determined that I would give my very best effort to follow them and hopefully earn as much respect as they did.

Discipline was a major factor in submarine school. While I was there I saw several men kicked out. It wasn't only that they couldn't cut the learning aspects of school, but they were also losing their tempers, arguing, and being just plain disobedient. That just wasn't going to fly. We were being prepared for the life we would face if we were assigned to a *boat.* Life on board a submarine leaves no margin for error. Every life depends on every other life and what we do every minute of the day or night can either save lives, or it can tragically cost lives; even the lives of the entire crew.

What each man does can and will affect the safety and well-being of all the men on the boat. There can be no mistakes made, especially due to a poor attitude or bad temper. "All-for-one-and-one-for-all" is an absolute, on board.

Any man will quickly and without hesitation, give his life for the good of others, and those others will all grieve in deep appreciation and honor toward any man who does.

It's an unspoken code of acceptance that we were made aware of right there in sub-school. If an emergency ever arose where a man must close a hatch door, locking someone behind it, so the crew and the boat would survive, it must be done. And it is very seriously accepted by every man, who with pride and honor, becomes part of serving in submarines. May God bless every one of them. Pride runs deep.

I GOT A BOAT

Time passed rather quickly at submarine school, there in Connecticut. Maybe it was because first of all, I enjoyed it, and secondly, I worked hard and applied myself. Thankfully, I was rewarded by graduating with very high marks. As soon as I graduated I was assigned to the USS Bass, SS K-2, a submarine stationed in Pearl Harbor. I was thrilled beyond what words could describe.

Photo Credit: U.S. Government/Public Domain
https://commons.wikimedia.org/wiki/File:USS_Bass_(SSK-2
).jpg

There were only three "K boats" in the fleet and they were all considered *sister ships* to each other. Dutch had been on the USS Barracuda, SS K-1 and here I was on the USS Bass, SS K-2. There was a third one, the USS Bonita, SS K-3. I was so excited, just looking forward to being aboard that I was like a kid waiting for Christmas.

Photo Credit: Hamilton 1:1 Communications, LLC;
USS Bass & Bonita, SS K-2 & K-3
https://designed4submariners.com/Barracuda_Class__SSK_.html

The USS Bass, SS K-2, a Barracuda-class submarine, was the second submarine of the United States Navy to be named for the Bass, which is an edible, spiny-finned fish.

The original Bass was commissioned in 1924 and took part in four war patrols during World War Two. Then, in 1945, after the war was over, she was *scuttled* (or sunk) and she was used as target practice for torpedoes from our own submarines.

The Bass I was on was originally named SS K-2. She was built at Mare Island Naval Shipyard, in

Vallejo, California. Mare Island was the first, and largest naval shipyard on the West coast, and evolved into one of the busiest shipyards in the world. More than 500 naval vessels were constructed there and thousands more were overhauled before the yard closed in 1996.

THE BUILDING OF MARE ISLAND

Many years before, it was Commander David Farragut, (born near Knoxville, Tennessee), who oversaw the construction of a new naval base at Mare Island and served as its commanding officer from September 16, 1854, to 1858. He was the Flag Officer of the United States Navy during the American Civil War and has the distinction of being the first Rear Admiral, the first Vice Admiral, and the first Full Admiral in the U.S. Navy.

His name has become famous for what he said during the Civil War at the Battle of Mobile Bay, at Mobile, Alabama on August 5, 1864. Farragut was in charge of several ships trying to take the port at Mobile and he was determined to take it.

Back then, there were floating wooden barrels of explosives in the water, which today we call mines, but in those days they were called *torpedoes*. They were put there to explode when a ship touched them. From that came the famous statement, "Damn the torpedoes, full speed ahead!" which has been repeated millions of times during various dangerous missions.

While Commander Farragut was trying to get through that harbor filled with those explosives, he gave the order to move ahead. That meant he was not going to fear the hidden torpedoes scattered throughout the water. He was going to disregard the minefield laid down around them and was determined to take the fort and gain control of the harbor. It was then that he was quoted despite being forewarned, saying, "Damn the torpedoes, full speed ahead!"

Farragut's victory at Mobile Bay cut off one of the last of the Confederate's deepwater ports and helped to secure President Abraham Lincoln's reelection in November that year. That's the man the United States trusted and called, ten years before, to build the Mare Island Naval Shipyard at Vallejo. I'm surprised they didn't name it after him.

ON BOARD THE K-2

The USS Bass, K-2--the boat I was now assigned to--was built at Mare Island, launched on May 2, 1951, commissioned on November 16, 1951, and put out to sea. She joined Submarine Division 72 at Pearl Harbor, arriving there on May 23, 1952, to join the United States Pacific Seventh fleet.

I thought, "How appropriate that I should be assigned to this particular boat, since November 18 is my birthday and she was commissioned on November 16. It wasn't exactly the same date or the same year, but close enough."

The three SS K boats, Barracuda (SS K-1), Bass (SS K-2), and Bonita (SS K-3), were equipped with large bow-mounted sonar systems as part of *Project Kayo.* These boats used experimental passive acoustics with low-frequency sonar, mounted on the submarine's bow. (See the previous photo.)

When the boat was rigged for silent running, these sonar systems offered greatly improved detection capabilities in locating other submarines in the area that were snorkeling or trying to hide.

Their main objective, because of their precise and highly advanced sonar power, was to hunt down and destroy other submarines. Even when those other submarines thought they were safe, whether snorkeling or fully submerged, they were discovered and wiped out. These boats became known as "K-boats", or hunter/killer submarines.

The success of this state-of-the-art sonar technology on the K-boats was eventually used in the development of nuclear-powered subs.

Getting assigned to the K-2 made me feel like I was following in the footsteps of my brother-in-law, Dutch. As I said, he had been in Pearl Harbor on the K-1, a sister ship to my boat. And to complete that picture, he was a cook and so was I.

Photo Credit: Scott Koen; The Stickleback
https://www.defense.gov/Multimedia/Experience/USS-New-York-A-City-on-the-Sea/

WHERE DO I BELONG?

Was I where I was supposed to be? I don't know, but life did go on from here because my time on the K-2 didn't last long. For some unknown reason, I never developed that feeling of *being attached* to the K-boat like I thought I would. Soon, I requested a transfer to the USS Stickleback, SS 415. This boat was headed for an extended cruise into the far northern parts of the Pacific.

Why I did it I'm not sure, but as they say, I was *young and dumb*--not that it was a dumb move for me to transfer--but I didn't have any grounds for that decision except that the opportunity was there and I jumped at it without any particular reason or motivation.

My transfer came through within days. It was signed and sealed; I was going to serve on the Stickleback. It's a decision I've never been sorry for making.

And I suppose it was just as well . . . because on June 26, 1957, the K-2 returned to the United States to operate along the west coast until about four months later, when on October 1, 1957, when she was de-commissioned and was put in reserve. That's the same as saying she was *moth-balled*.

CHAPTER FOUR

SEA DUTY

The Stickleback had just come through a major overhaul at the Mare Island Naval Shipyard, lasting from September 30, 1955, to February 10, 1956. From there she was moved to Pearl Harbor where I transferred to her.

She was a guppy-snorkel-fast-attack boat and for the first month, we went through some intensive training in the Pacific Ocean near Oahu, Hawaii. Every system was checked out. We also played a couple of war games with some surface ships to see how efficient we were.

Every man on board was getting used to working with, and trusting in, each other. I could literally feel the intensity in the air as we worked one shift after another and then began all over again with hardly a break in between. The whole crew seemed to be charged with electric anticipation.

I had been on board only a short time when just as I had been told, we left Honolulu and headed out to sea. I was informed that we were heading to Yokosuka, Japan, (pronounced Yō-koos-ka). The thrill of actually being fully out to sea as part of a submarine crew was overwhelming to me. And it was real. I was in for an amazing ride. This is what I had been waiting for.

But then I began to hear that all was not going to go as well as we had hoped. There was a problem with the boat. There was concern about our port (left side) shaft; it was squeaking.

I LEARNED BY LISTENING

The cruise was going to take us out into the Pacific Ocean and far into the northern hemisphere.

And from the talk I was able to pick up here and there, I understood it was not a good idea to be making noises on the way to where we were going. I gathered from some of the older men who had been there before, that wherever this was, we needed to be extremely quiet for this mission. A noisy shaft would have canceled it.

Nevertheless, on March 9, 1956, we left the Hawaiian Island area and headed for Yokosuka, which was to be our first stop within many thousands of miles and many weeks at sea. I had no idea what life was all about but I was glad I made the change to the Stickleback and was on my way to something exciting. At least I hoped it would be.

Photo Credit White Hat Bar
http://graz-web.com/Graz/Japan/
Yokosuka-Honcho.htm

A STRANGE NEW WORLD

We got into the port of Yokosuka late in the afternoon and of course, it was all strange to me. The wildlife in this foreign country was just about to turn me upside down. I didn't smoke and I didn't drink, but I was beginning to learn how to do both. Up until that point in time, I didn't even drink coffee. So, I was green, I was young, I was innocent, and this was no longer a place where I could lay beneath a shade tree to read a book and enjoy life. This was a different world.

The nightlife was wide open with drinking, drugs, prostitution, and an array of black-market items of any sort you could imagine. This was not a place for a young, innocent, midwest boy like me. There I was, young and dumb, and right in the middle of a life I didn't understand at all.

After carrying on half the night, I woke up the next morning in a room on the third floor of a hotel, having no idea where I was or how I got there.

GONE!

My first thought was that I was in deep trouble for not getting back to my boat the night before. (And of course, I was in trouble.)

I hurried down the stairs and caught a cab back to the base, to the docks where the Stickleback was supposed to be, and . . . *it was gone!* That's right. *Gone!!* The 415 was nowhere in sight. Now, what was I supposed to do?

The Korean war was just beginning to wind down and we were deep into a cold war with Russia. Knowing that I had *missed movement* of my boat would leave me wide open to receive a *court-martial.* I guess since this was still during wartime, I could even be tried for desertion! With all that running through my confused and frightened mind my worries multiplied.

"Wow. Could I even be shot?" I didn't know. What a situation to be in, for an innocent, frightened 18-year-old kid.

BY THE GOODNESS OF GOD

I took a cab to the base and once I arrived, I asked the Japanese cab driver if he knew where the office for the Captain of the base was located. He said he did, so I told him to take me there. I had no recourse other than to turn myself in and take my medicine.

As I tried to explain my situation to the cab driver, he began to explain in broken English that I didn't have to go to the office of the Base Commander because he knew where my boat was. I was dumbfounded. I didn't think he was able to really understand my English or knew what I was talking about.

I said, "Really? Are you kidding me?"

No, he wasn't. He knew exactly where my boat was and he could take me to it. I was amazed! I breathed a sigh of relief. Although I knew I wasn't entirely clear of trouble, at least it wouldn't be as bad as I thought.

He took me through some left and right turns, here and there, and suddenly he pointed to a submarine completely out of the water, sitting in what is called *dry dock.* Sure enough; it was my boat. When I saw that big "415" painted on the side of the *sail*, I knew God was being extra good to me. I also knew I didn't deserve it.

IN TROUBLE

Fear, frustration, and confusion were rushing through my being. What should I do? What *could* I do? Where could I go to turn myself in? My boat was there, yes, but it was sitting high and dry instead of in the water, with no apparent way for me to go aboard.

There was nothing for me to do but pay the driver and try to figure out my next move. At least my boat was here and I had not *missed movement.* I was thankful for that but knew I was still in deep trouble with plenty of problems in front of me. There would probably be a *face-to-face* with the Chief of the boat, (COB) first. Then my Division Officer would confront me, and eventually the Captain. Yeah, I was in deep trouble.

I asked questions of the workers at the dry dock and I was directed as to how I could get back on my vessel. I walked the *gangplank* to go aboard, and one of my shipmates, the watchman on duty at that time, looked at me wide-eyed. He slowly shook his head from side to side saying, "Man you are in deep *do-do* all the way up to your neck!"

Now I was getting scared. Since that was *his* attitude I couldn't begin to imagine just how bad it would be when I had to face the Captain. But . . . I would soon find out.

THE COB

The deck watch told me the Chief of the boat happened to be on the *after-quarter-deck* right then, so I walked back there and reported to him first, just to get the ball rolling. But I wasn't prepared for his attitude. He looked at me with a deep, growling frown, stared me straight in the eyes, and tried to speak . . . but he couldn't.

Suddenly, he broke into a grin, turned his back on the other people there, shook his head, (and quietly) laughed, deep in his chest. They couldn't hear him, but I sure did.

I thought, "Hey man, this might be funny to you but it sure isn't to me. I'm in trouble here!"

He said, "Come on son, we'll go see *The Ole' Man* (as the Captain is affectionately called) and let him know you're back on board and that you're alright. You've put a *** of a scare into everyone. We've all been worried about you."

As much as I appreciated his lighthearted friendliness, and the fact that the people were worried about me, that should have made me feel good, putting me somewhat at ease. But no. I still didn't feel too great. I had messed up, I knew it, and I was scared.

THE SQUEAK

On our way back through the base, the COB told me that our boat had been moved to dry dock yesterday because we had a serious *squeak* in one of the propeller shafts which we knew we had ever since we left Honolulu.

Then he said, "Since we're eventually going to be heading into some *tough areas,* we would easily be picked up on sonar and . . ." And that's where he quit talking.

After a long and seemingly forever-lasting silence he said, "Well anyway, the shaft had to be fixed so we'll be here for a few more days. Meanwhile, we'll see what the *Ole' Man* wants to say to you. And you know that was a stupid thing you did, don't you?" Yeah, I sure did.

THE OLE' MAN

With me walking quietly beside him, the COB found his way to the building on the base where the Captain was and we went in. There was a young

Marine on guard duty at the entrance and the COB told him why we were there. The young Marine saluted the Chief and told us to come in and wait. Then he turned sharply, and walked across the room, knocked on a door, and went into an office. Within only a minute or so he came back out and told the COB that the Captain wanted us to go back to the boat and he would see us later on that day.

I thought, "Well, here it is. Now I'll be kicked out of the Navy. Why did I let such a stupid thing happen?"

So we turned around and walked back to the boat; the trek was long and quiet, even though the COB was talking the whole time. I heard his voice but I did not catch an understandable word because my mind was elsewhere. I was really worried.

TEMPORARY INSANITY

I don't remember too much about all the remarks made to me by my shipmates or even how many days it was before my Division Officer, Dave, told me he wanted some time with me to put together my defense.

"My defense?" When I heard him say that I knew my goose was cooked. He had only been in the Navy a couple of years himself and this was also his first sea duty. I was surprised at how quickly we bonded; maybe it was because we had that in common.

I thought back to when I was offered OCS out of boot camp. I would have been an Officer by this time, too. But no . . . I had to mess everything up.

Acting like we were *old friends,* we even made a few jokes about my situation. At least I thought they were jokes. Dave said, "I'm going to plead *temporary insanity* for you and by doing so, maybe we can get you off. Or at least you might not be punished too severely. That way, whatever you get, it will probably be less than two years."

I thought he was kidding about the insanity plea, but when he said two years I panicked. He was an Officer and I knew better than to balk, but my survival instincts kicked in and I bowed up and said, "What? Two years? Are you nuts?!" I felt my sense of panic rise. "There's no way you're serious, are you?"

No. He wasn't serious. Well, not entirely. Thankfully he told me he was just kidding about the two years, but he did want to make a plea of temporary insanity since I had never drank before. He wanted to blame it on the effects of the alcohol and how it messed me up beyond what was *normal* for me. I didn't know what else to do so I just kind of nodded in approval.

TAKE IT LIKE A MAN

I felt better knowing he was kidding but I told him I couldn't have him present an insanity plea. That would probably get me kicked out of submarine duty and I didn't want to lose that. Even so, I still knew I was in a serious pickle with the Captain and was hoping there was some way Dave could get me out of it.

But temporary insanity? No way. I would just face it with the truth and take my licking like I needed to. I would take whatever it was, but I didn't want to be kicked out of the Navy.

I hate to say this but I can't remember my Division Officer's full name. I do remember that his name was Dave, and even though we became fast friends, I don't remember his last name anymore. It's been many years since we've seen each other and we will probably never meet again.

Well, we did meet up just one time and I'll never forget the way it happened.

CHAPTER FIVE

THE PAST CATCHES UP TO ME

Several years had passed since my discharge and I was managing a gas station in Dallas, Texas, near the LBJ freeway. There was a restaurant nearby where I generally ate lunch and that's where it took place.

I was sitting at my table eating and I thought I heard someone say "Hey Veer. You dumb *blankety-blank!"*

Immediately I looked up, scrutinizing each table near me but I saw no one looking at me. I guessed that I was mistaken and went back to eating my lunch.

Then I heard it again, "Hey Veer. You dumb *blankety-blank!"* Once more I looked around and saw no one that I knew.

When I heard it the third time I was sure of what I heard, so I kept my head bowed down as if I was looking at the food on my plate, but I was looking upward through strained eyes, scanning everyone in the room.

Then it happened a fourth time but this time I saw the guy who said it. He was sitting at a table across the aisle about seven or eight feet away from me. That was it. I had had enough of him, whoever he was!

I got up and walked over to him ready for whatever would happen next. I got almost to the table, and before I could speak, he broke into a big grin and said,

"Been in any good Marine brigs lately?"

Boy, was I ever shocked. There he was, smirking at me. It was my Division Officer, Dave, who got the Ole' Man to go easy on me years before, at my Captain's

Mast, (the non-judicial punishment of an enlisted person by a military commanding officer).

MRS. CRUISE

Yes, I stood *Captains Mast* and the Ole' Man dressed me down good. But in a sense, the way he treated me made me feel the same way I did when I was a child. I was on my way to school one day but turned around to go back home and play.

Later that day, I was taken before Mrs. Cruise, the Principle. I told her how my brother had been teasing me about a girl named Anna Lea Foshua. (I hope that spelling is close to being correct.) I wanted to dodge the embarrassment of the rumors and all.

Mrs. Cruise dressed me down good but in a sort of motherly fashion, remarking on my excuse, "Oh, come now young man. You're just talking through your shoestrings." She let me go to my class with a kindly lecture. No harm done.

Well, the Captain was no school Principle for sure. And this was far more serious than my being late to class. Even though he sentenced me to serve a month in the *brig* on the Marine base in Yokosuka, nevertheless, there was a kindness about him; almost a fatherliness.

No, being in the brig wasn't any fun, especially with me being a Sailor instead of a Marine. And to make it even worse I was the *only* sailor there! But the truth is, the Ole' Man could have been even more severe if he'd wanted to. It was within his rights and within his means to do so. Even today, I still have very good memories of him.

TOUGHING IT OUT

One of the things the other prisoners and I had to do while I was there was break up an old airport runway with nothing but sledgehammers and pickaxes.

It was extremely hard work, but I managed to tough it out. I went there as a young guy weighing in at about 135 pounds. I came out at about 150 pounds having packed on some very hard muscle. Not that I wasn't in good shape when I started my sentence but all the physical labor of swinging a pickaxe and sledgehammer all day had a telling effect.

I'll always be thankful that I survived and made it back on board before we went back out to sea to complete our mission.

A CARING CAPTAIN

It had taken many days for workmen to repair the noisy shaft but it was actually to my benefit because it allowed me to serve my time in the brig and get back before we put out to sea again.

Now, if I remember correctly, I never actually served my complete sentence in the brig. When we were ready to ship out the Ole' Man sent for me to be released before my full time was up.

Once more he called for me to see him in private and he talked to me like my mother would have. He said that since I didn't actually *miss movement* in the strictest sense but was AOL, (late coming back from liberty), I needed to have some sort of disciplinary action to let me know that rules were meant to be kept, not broken, because the whole crew had to depend on one another.

Then he sent me back to work with the promise that I would do much better or there would be no reprieve if there ever was a next time. Oh, I sure was blessed with a good Captain. He was as good as a mama to me and it was good to be back on board again.

SIGN YOUR NAME RIGHT HERE

At that time, the few submarines based in Japan and Pearl Harbor were all rigged with a full war load of torpedoes, and they were being deployed to the waters neighboring Korea to look for Soviet ship movements. Also, I had changed my rate from commissary to torpedo man, so I was involved with fully loading the boat.

Unlike World War Two, when *seek and sink* was the order of the day, during this era, the submarines of the Far East Command were instructed to *look but don't touch;* then report the findings and wait for further orders.

At this particular time, the Navy Department was very concerned about the 50 or 60 Soviet submarines that were reported to be in Vladivostok, Russia, and what their movements were. The Korean War was winding down, but still active, so the mission for U.S. submarines became the surveillance of Soviet and Chinese shipping in the East China Sea and the Sea of Japan.

Because of the classified nature of this assignment, each man on board had to sign a document swearing that he would never discuss the details of where we had gone or what we had seen or done once we got there. If we ever violated that sworn oath, even if we were discharged and back into civilian life, the punishment would be an arrest and a sentence of unspecified time in some facility such as Leavenworth Military Prison in Kansas. No, I didn't want that.

For that reason, I had originally decided to omit dates, locations, and much of the information I had sworn to protect with my silence. Many of these details now exist in the public domain; however, I have been in touch with Government agencies such as the Navy Office of Information, the Deputy of Public Affairs office, and the Defense Office of Pre-publication Review. Since they have declassified the information I am able to use it here. Nevertheless, I have changed some of the names and locations of where we went and what we did thereby holding true to the document we all signed.

Even so, I'll try to share some of the stories about our crew and what I enjoyed for the next couple of months, though enjoyment wasn't exactly why we were there.

WHAT WILL HE DO?

Finally, my brig time was behind me, thanks to the graciousness of the Captain. Now the time had come to leave Yokosuka and head out to sea. This was where we would see if the work done on the shaft while we were in dry dock was successful, making our operation quiet . . . It wasn't.

We still had some noise coming from the port shaft. But the Ole' Man decided to give it a try and head out into the deep waters of our northern trip. If he was satisfied that we could manage with a small amount of noise from the shaft, then we would continue onward to accomplish our mission. But would he be satisfied

We knew he had a plan, but we didn't know what. Someone said he was going to give it a try by running with our own ships first. We didn't know then, but the Ole' Man was pretty good at keeping us all informed, when possible. So, we just had to wait, knowing that when he was ready to tell us, that's when we'd know; not sooner, and not later.

There was a *task force* of U.S. ships nearby consisting of an aircraft carrier, the USS Yorktown, (recently re-commissioned two years earlier), changing it from an attack carrier, CVA 10, into an antisubmarine carrier, CVS 10. And there were six destroyers, two destroyer escorts, and a light cruiser.

Taking the opportunity to play some war games with them, the Ole' Man decided to spend a day or two checking out our screws to see how easily another ship could detect the noise from our repaired shaft.

Part of the war games consisted of carrying out some fake torpedo runs on the other ships, knowing they would be listening for us. By doing this, the Captain would decide if the squeak in the shaft was, or was not, detectable enough to keep us from making the northern run that had been assigned to us. We trusted in him completely.

FIRE ON HER?

It was three-thirty in the morning and we had been following various ships of the *task force* all night. We were playing *cat and mouse* with one of the destroyers when the Yorktown suddenly turned in our direction like they had picked us up on their sonar. So, we went deeper as Captain Denny passed the word over the intercom: "Rig for silent running".

Sitting as quiet as we could, we listened to the Yorktown pass over us. Then we began following her, hiding in the cavitation (noise) of her screws churning up the water behind her. I guess we followed close behind her for at least a half-hour when the Captain passed the word that we were going to fire on her.

I thought, "Fire on her! Is he crazy? She's one of our own!"

Well, I was still young, new, and not familiar with how war games were played between our ships. That's when, from the other more seasoned men, I learned that was exactly what we were doing; still playing war games, and it was just a figure of speech used to describe what we would do next.

Hidden in the noise of her screws, we continued to follow behind the Yorktown, getting closer and closer, until we were practically underneath her. That's when Captain Denny decided to fire a flare into the air behind her. As it broke the sky it was an indicator to the Yorktown that had it been a torpedo we would have hit her.

We all cheered when the Captain said he could see the flare in the air behind her. Then he sounded the alarm to come to the surface, showing the Captain of the Yorktown just how close we were to her.

The Yorktown Captain didn't like it but we showed him that we, the crew aboard the USS Stickleback, were ready for war if it became necessary. We had proven to ourselves that even with the shaft not as quiet as it should be we could continue on our way farther north to complete our assignment.

HEADING BACK

With that little episode over, we headed back to Yokosuka and I had a very sobering thought. "The next time this happened we might not be playing games with one of our ships and the flare could be a real torpedo."

Honestly though, in the depth of my heart, I hoped there wouldn't have to be a *next time* if that would be the end result.

The war games were over and the checkups were all done. I got a chill from head to toe as I realized that now it was time for the real stuff to begin.

Despite the noise coming from our port shaft, the Ole' Man decided to leave Yokosuka and take the Stickleback far into the North Pacific and beyond. That's why we were sent here and it was time to do what we were supposed to do.

This young sailor boy had a strange chill down his back. I realized just how serious all of this was. We were here. I was here. This was real. And it was absolutely time to grow up. "Welcome to the real world, son."

At the same time, there was a thrill I'll never be able to explain. I guess you'd have to *be there* to understand what I mean.

SONAR HEARS EXPLOSIONS

It was the middle of May 1956, early in the morning before sunrise. We quietly slipped out of our berth at the Yokosuka docks and headed deep into the North Pacific Ocean to parts yet unknown to us. But our travels were interrupted. The sonar man began to pick up multiple underwater explosions which he reported to the Captain.

After tracing them carefully all day we finally sighted a Soviet Hydrographic Research ship and a Russian destroyer. Putting two and two together, it was decided that the explosions we heard were coming from hydrographic experiments being conducted by the Soviet ships.

Unfortunately, our squeaky shaft kept us as far away from them and as quiet as possible, which limited our effectiveness in gathering information as we had hoped.

The Captain decided to cut our trip short. With only fourteen days in the patrol area, we turned around and headed back to Japan. It was the beginning of June and we still had time on our intended cruise, so instead of returning to Pearl Harbor, we stayed in Yokosuka until the end of June.

SHOE SHINE?

I guess the Captain decided we all needed some rest and recreation so we set out for Hong Kong, China, to relax and enjoy our leisure time. We were to take a six-day break from the rigors of the patrol.

Before we went ashore for liberty, one of the officers had a sit-down talk with us. He warned us about becoming sexually involved with the *ladies* in the bars and told us all about sexually transmitted diseases. There was a lot of that being passed on to servicemen. But one of the strangest bits of information he gave us was how to interact with young boys offering us a shoe shine.

He told us how these young boys would come up and offer to clean any stains on our uniforms, but we had no stains; at least not until he pointed out something quite odd. They would have something in their hand and would touch us with it. It would then appear on our blue uniform as a white blotch. It looked just as white as flour, but we wouldn't be able to brush it off. For a price, however, they could fix it.

They would also touch our black dress shoes, telling us they needed to be cleaned, and yes, the blotch was there too, and you couldn't get it off. He told us it was a major scam to get money and it would make us mad, but . . . we shouldn't dare touch these people or push them away.

He said, "The minute you put a hand on one of them the Hong Kong Police will arrest you for physical abuse. So *do not* touch them. Instead, kick them with your foot and you will not be bothered by the police." Yeah, I had a hard time understanding that.

He continued, "But if they do happen to stain your uniform or shoes the best thing to do is to simply ask them to clean it for you and then give them a couple of coins. They will go away looking for the next guy and you can go and enjoy yourself."

So we went on our first night of liberty in Hong Kong. Thankfully, I was never bothered by any of those boys.

HE'S DROWNING!

Two days before the fourth of July, we left Hong Kong, and went to Taiwan, in the Republic of China, where we once again operated with a *task force* of our own ships for about a week.

Taiwan is a simple island with many small villages strung together from end to end across its territory. It is a very poor country and so we didn't get any leave to go ashore there.

But while we were in the harbor, boys, very young boys, swam to the side of the boat asking for money. Several of the guys threw money to them and they were quick to grab it. Then one of the guys threw a coin too far from a boy and he dove down under the water after it. It seemed like he would never come up and the guy that threw it was afraid he was drowning, so, he went in after him.

We found out later that the boys are very good swimmers and they would often dive for coins that were thrown to them. Anyway, I was proud of that shipmate for diving in after that boy.

We left there to join up with a different *task force* and operated with them until the end of July.

The shaft was still noisy so once again we returned to Yokosuka, hoping to quiet it down even more. We had a short routine maintenance and repair period from the end of July to the beginning of August, still hoping to fix our little problem

CHAPTER SIX

GEORGIE BOY

During that short stay in Yokosuka, my friend, Mac Ordonio (I'll tell you about him later) and I walked into a nightclub one evening and found a place to sit, at the bar. After looking around for a few minutes I noticed three Marines sitting at a booth right near us. I could only partially see the one who had his back to me but I thought I recognized him. My first thought was that he'd been one of the Marine guards at the brig I spent time in.

Taking a chance it might be him, I got off my bar stool and walked over to him, tapped him on his left shoulder, and in a friendly manner I said, "Hey, do you remember me?"

I didn't expect what happened next. He turned and looked at me with a frown and a tough "don't bother me" look. Then his eyes widened and to my surprise, he said, "Hey, wait a minute. No way, Veer. No way! You're not still mad after all these years are you? I mean that was a long time ago, right?"

I had no idea what he was talking about. Then I recognized him as George, a boy I went to school with. He used to pick on my sister, Rosie. Each time she told me about it I would *lay in wait* for him after school. Several of those times I managed to catch him and beat him up but when he picked himself up he would run away and call me a name. He was much faster than me so of course, when he ran away I couldn't ever catch up to him. And so, his menacing continued until the next time I saw him.

Later on, I think he and his family moved away. And now, many years had passed and I had forgotten all about him until we were face to face again.

I said, "No. Not at all! You're right Georgie boy. That was years ago and we were just kids so it's all behind us. We're grown up now and the past is the past. It's good to see you again."

He stood up to face me and offered to shake my hand while asking, "So there are no hard feelings about those bygone days when I acted so stupid, right?"

I graciously replied, "Not at all, George. Those days are over and I'm glad to see you. *Semper Fi* to you, man. You all stay safe and get back home safe, Okay?"

I must have had some wisdom somewhere inside of me as I was able to keep things civil. That was a good thing because as he stood up it was very apparent he was three to four inches taller than me and had me by at least 40 pounds. *Little Georgie* had grown into *George*, a full-size Marine. Yeah, I *was* glad to forget the whole issue!

We shook hands and I said, "I hope to see you back in Freeport someday. Stay safe."

Then we each turned back to our tables to enjoy our company.

Who knows where you might meet up with your past? It's a good idea to keep it simple. Besides, kindness never gave anyone a black eye or a bloody nose. At least it didn't this time.

ROUTINE?

It was now the middle of August and we continued playing war games with our fleet in the Sagami Bay area of Japan. The bay is about 25 miles south of Tokyo, in the Hanshu/Kyoto area. We weren't there very long because the Captain felt we had to head back to Yokosuka to get yet *another routine* maintenance checkup, (so we were told). But we all knew it was more than that.

We knew it was because the Captain was concerned about continuing onward to our designated assignment with a noisy boat. He was a man that took everything as it came and did the best that anyone could do. He was concerned about taking a chance with that shaft, and we understood him.

I can't impress upon you enough just how important this was. Because of where we were going, it *had* to be quiet.

Finally, they were done checking the boat and it was time to leave Yokosuka behind and head North. We didn't know where *just North* was but we all had the feeling *this was it;* it was time to go.

ANCHORS AWAY

Once we were a couple of days out to sea, the Captain came on the intercom and announced he was going to reveal the orders for this mission. Up until then, the orders had been sealed so even the Captain didn't know exactly what was involved. We were all excited and nervous to find out where we were going and what we were going to do.

Within a few minutes, the Captain was back on the speaker and read the information to the crew. The orders were that the Stickleback and her crew were to head out for another cold war run into the far northern parts of the Pacific, taking us into the Bering Straits near Siberia, Russia; noisy shaft or not.

The tension in the boat was so thick you could cut it with a knife. But it wasn't a fearful tension. Every man on board was confident that he knew his job, and that he was going to do that job extremely well. He was also assured that every man on board *with* him knew his job too, and was going to perform it with nothing less than excellence.

Besides that, everyone trusted the Ole' Man completely. On submarines, that trust and confidence in one another is what makes us all tick.

WE ARRIVED

A week later, we arrived at our assigned patrol area, in the shipping lanes off the coast of the Soviet Union, only 50 miles across the water from Alaska. The shoreline there is called a *shelf.* Although it's about 300 feet deep, it extends outward into the sea by just roughly 40 feet in width.

Most ships try to stay outside of those 40 feet, making it sort of a boundary to be respected. But the 300-foot depth makes it deep enough for a submarine to lay quietly and hide if necessary. It could certainly work to our advantage. We were here, but we certainly didn't want it to be discovered.

I was filled with a dozen different mixed emotions, ranging from excitement over seeing new places, to a crazy kind of fear over what could happen to us. But, I was able to put that fear away, knowing we had a mission to complete, and I could rely on every man there.

And *every* man meant from the *mess boy* (kitchen help) to the Captain. We were all in the same boat, (no pun intended), and we would make sure we all made it home together.

I was definitely changing. I was becoming someone better than myself. This young, innocent, teenage boy from the midwest was excited to be doing something new and important for our country.

Here I was in a foreign ocean, near a foreign land in the middle of nowhere. It was all much grander than anything I had ever done in my life, or perhaps would ever be privileged to do again.

I knew this wasn't *the other side* of the *big blue ball* I dreamed of as a boy. This was a long way from there. But to be completely honest, I felt a little homesick. So, I was glad that I was surrounded by so many good people.

CONTACT BEARING

I was now extremely comfortable living in close quarters with my shipmates and I had become very good at my duties: spotting lights on the horizon at night from other ships. But it was easy to get lost in the wonder of it all.

For a young, innocent eighteen/nineteen-year-old I was living a fascinating and amazing dream. This wasn't make-believe. I was getting a glimpse of the outside world.

I had to remind myself about the where and the why of my mission. I'd have to shake my head and realize this wasn't a fantasy. This was true in every sense. The ocean, the darkness, the reflection of the moon over a distant sea. All were very, very real. Yes, Dad, *life is real.*

I, like everyone else here, had a very serious job to do and we needed to give it our very best and our full attention. Every man was dedicated; this is what we volunteered for, and this is what we trained for.

Now, amid the excitement, the newness, and the ever-present possible danger, I searched for that inner assurance that we would get through the challenges ahead and so we could make it back home safely.

So, standing there on the *bridge* I bowed my head for a minute and prayed. I asked God to keep us all safe. It was natural to be nervous way out here but I knew God loved us enough that He wanted us to be kept safe, too. I believed He would do that for us, and I relaxed.

I turned my attention back to my job, diligently searching for any small glow of light beyond the horizon caused by a contact. And I wanted none to be seen; I wanted us to remain *unseen* and out of harm's way.

ENJOYING LIFE ON BOARD

I had come to absolutely love submarine life. I served with some fine men and we appreciated each other's company. We also knew how to have fun. There were table games that went on almost all the time. Games like Acey Deucy, Dominoes, Cribbage, Poker, and a few others were played with either cards or dice. The very regulated routine we lived by became just as natural to me as my life was before I ever dreamed of joining the Navy.

I also loved the jukebox-like music we had at the mess tables. They were just like the ones found in restaurants that you put money in to play a record. The difference here was we didn't have to put money in.

Even standing duty on the bridge at night as a *lookout* gave me a sense of being part of something marvelous and unique. I came to feel so attached to it all, I don't even know how to share or describe it. But without a doubt, these were some of my most enjoyable times.

UNEXPECTED BEAUTY

From the bridge, I could see the bow of the boat breaking through the murky sea. It caused a curling pattern of water that went upward, outward, and away from the knife-like sharpness of the bow, cutting through the darkness of the ocean. It created a magical light show caused by the disturbance of the plankton as it churned and blended with the sea, turning it into a bright, mesmerizing, cascading glow of dancing waters that lighted the dusk around it.

About twenty percent of marine species are bioluminescent, meaning that when they are disturbed at night they emit anything from a soft warm glow to a brilliant glow like we used to see as kids in the fireflies of the night.

Standing on the boat in the wee dark hours I saw this beautiful glowing display. A man could get lost in the wonder of it all, and though it drew us in by its beauty, there was always the dire importance of the job in the back of our mind. That kept us on our toes and ever alert and watchful for what might be over the horizon where the colorless sky met the black waters below it.

That picture of the ocean that was so alive, and the memory of those hours on the bridge, will be with me forever, as long as I live. I feel I was secretly and quietly given a private sneak preview of the hand of God at work. It was amazing.

SWIM CALL

Sometimes during the day, when we were much further from land and all sonar and radar reports indicated there was no present danger nearby, the Captain would call for some recreation time; just something to help us relax a little. The boat would surface and come to a place where the men could take a break *topside* for some coffee, sandwiches, or soft baked sweet rolls and fresh air.

At those times, which were few, the Ole' Man would authorize a *swim call.* That's when dozens of crew members would head up top to catch some sun and hit the water. It was always enjoyable. We put all the tensions behind us and for a small space of time, we could make believe we were just a bunch of plain, young, goofy, happy-go-lucky guys again. With some extraordinary exceptions, of course.

As a team, we sailed beneath the waters in great iron ships and as often as we safely could, we needed time to breathe and put away all that stress and strain. Even if it was just for forty-five minutes to an hour, it was always worth it.

Taking a little time for enjoyment has a great strengthening and calming effect on all hands, including the Captain. After all, every one of us from the Captain on down is just flesh and blood.

WHEN DUTY CALLS

Watching for ship contacts was always extremely important so when the men were swimming in the water and were called back to duty, they'd have to get back onboard sometimes at a minute's notice. Any delay waiting for them to get below decks would hold us back from diving immediately. That could be disastrous.

When the crew does finally get a swim call there are always men standing by on the bridge, with rifles, watching for sharks. There is an extra lookout on the bridge as well, where they can get a good view in the distance of any possible sharks wanting to visit us and make a meal of us. They also search the horizon for any possible contact with other ships.

There's always a code of seriousness even at play but the guys on the boats need some lightheartedness, too.

And about the one keeping watch on the bridge, the joke is, "He's not there to shoot the shark--just the guy who is farthest out, so if he gets in trouble, the rest of us don't have to swim out there and get him".

Yeah, funny, right?

Try as we may, the mission at hand never does get too far away from our thoughts. That's the way it is with these men who go out to sea in these subs. They keep the conversations and attitudes light until the time comes to get down to serious business again. Then it's one for all and all for one, and they mean it with all their heart.

Any man serving on board a submarine knows that at any time, if the emergency arises, and it becomes necessary to have a hatch closed on him so that others can survive, he will give his own life to save the others: his crewmates and his brothers. Yes, he really would willingly die by that code if that time came. He may even be the one to *dog down* the hatch on himself.

CHAPTER SEVEN

ALL HANDS BELOW DECKS

All too soon, the enjoyment of the sun and fresh air is over. All hands are ordered to go below decks again where we go back to being very, very intent about our commitment. We might play for a little while but we never forget where we are and why we're there.

After a short time of relaxing, our minds are back on our jobs, knowing that we're just off the Siberian coast collecting information on the movement of Soviet ships. With that realization, *playtime* is over; it's washed away with the wet and the waves we were just unwinding in.

Once below decks, every man returns to his duty station, and back to the tasks at hand. The torpedo man goes to either the forward or the after torpedo room, the cooks go to the galley, the radioman to the radio shack, the engineers to the engine rooms, and the Captain goes back to being the most crucial man on board.

The Captain has the toughest job of all. He is the man who must shoulder the responsibility of keeping the mission going forward. Every decision he makes is with the idea that he is going to complete his mission and keep his men safe doing it.

CLEAR THE BRIDGE

One evening when swim call was over and darkness had fallen, we were up on the bridge for night lookout. Carefully scanning the distant horizon, one of the lookouts confirmed that lights were seen coming in our direction. Sonar had reported a possible contact but I guess the Officer of the bridge wanted to be sure of what we were seeing. Was it shore interference or was it a ship? If it was a ship contact then we needed to be below the surface, not riding on it in plain sight for whoever was out there.

Suddenly, the Officer of the bridge was sure that it was a legitimate contact and we needed to be hidden from their view. He yelled to us and the Conning Tower just below, "Lookouts, below! Clear the Bridge." And in the same breath, he hollered to the Control Room, "Dive! Dive! Flood negative. Take her down to periscope depth."

So the men on watch got us down as quickly as possible. There was no fear nor undue excitement because every man knew his job well enough to do it in his sleep. When the order was given everyone jumped into action. We all knew time was crucial in getting her off the horizon and below the surface; there was only one possibility for who could be out there. It had to be a Soviet ship, and they may have spotted us.

Even if we were in international waters, which is twelve miles off the coast at this point, we did not want anyone at all to know we were there. This was my first taste of true excitement and possible danger because of our location.

ORDERS FROM THE BRIDGE

The *negative tank* is a main ballast tank and all main ballast tanks have hydraulically operated vent valves that can be rigged for hand operation if necessary. By opening these vent valves the air in them escapes out of the top and the tank quickly fills with water from the bottom. This adds more weight to the boat. The faster it is done the faster the boat can submerge.

At that moment we were in a hurry. When the Officer of the bridge yelled, "Clear the bridge. Dive! Dive! Flood Negative! Take her down fast!" we knew we had to get completely under the surface of the water as quickly as possible. And we did.

Right on the heels of those orders came the Duty Officer's call to "rig for silent running". And it was a good thing he did. But then just a couple of minutes later he called, "Take her down to one-five-oh feet. Silent running is still in effect." From the periscope, he had spotted a couple of ships headed our way and being 150 feet beneath the surface would give us the protection we needed.

THAT DREADED SOUND

After a while we started hearing that dreaded "swish, swish, swish" from the cavitation of the screws on the ship that had been spotted. Yes, it was searching for us. And soon, it was joined by another ship and they continued back and forth over our position for a couple of hours.

Right after we were down and were positive there was a ship up there checking on us, the Ole' Man moved to the control room and called for the word to be passed, "*Mouth to ear,* to all compartments. Be extra sure we are *rigged for silent running.*" He didn't want so much as a pencil to drop on the deck floor because it would make enough noise to be picked up by their sonar. So, the crew of approximately 80 of us sat as still as we could be, each one being very, very quiet.

We had drilled many times for this but doing it for real was a bit unnerving. I had visions of our being depth charged and no one would have ever known what happened to us. I'm sure that was on the minds of many other guys, too.

First, ten minutes passed; then half an hour; then an hour. Still, we could hear the slow-moving "swish, swish, swish" continuously moving back and forth overhead. I guess they had a pretty good idea we were there but they just couldn't get a bearing on us. Thank God. Perhaps this was the Lord answering my earlier prayer from the bridge.

HOW MUCH LONGER?!

By the time we were down for two hours with everything turned off, the boat was getting pretty hot. The bulkheads (walls), the pipes, and any other metal surfaces were getting wet with condensation. We were also sweating and most of the men had taken off their shirts and some had even taken off their t-shirts.

Two and a half hours passed and the cavitation noise from the screws overhead seemed to have left us, so some of the men began whispering, wondering if we shouldn't go up and take a look around. But they weren't in charge.

Even though it meant more waiting, more sweating, more silence, and more worrying, the Ole' Man decided to wait another 30 minutes running silent, just to be safe. The more I got to be around him, and the more I got to see him in action, while under much stress, the more I appreciated him; even more than before. He was the *real deal.*

By this time, everyone was perspiring like crazy but the Captain was being careful and I was glad he was in charge of our safety.

I did feel sorry for the officers, though, who kept their pants, t-shirts, shoes, and socks on while many of the men were stripped down to nothing but their skivvies and were still dripping with sweat like crazy.

We had often been through the drills for this kind of situation, practicing our fake, make-believe, war games near Pearl Harbor. But this was real and the pressure could be felt by everyone. Still, I was very thankful for the many hours we had spent doing those training drills. They made us ready for such a time as this.

AT EASE, MEN

The noise from above us was gone and we heard the order from the Captain, "Come to periscope depth." He got the scope up, had a look around, and everything seemed to be quiet. Still, we remained at periscope depth and rigged for silence until the Captain decided to take a chance and surface again.

So, after a short while at periscope depth, he very carefully brought us to the surface, but this time it was only with the decks awash, barely covered with water, and still using the periscope. Then I guess he must have been satisfied that we were safe because he turned it over to the Officer on duty, who called for the lookouts to go back to the bridge.

OPTIMISTICALLY CAUTIOUS

It was good to be topside again and into some cool night air, but we were ten or twenty times more cautious about looking for contacts. They were out there somewhere and we sure didn't want them to know that we were too.

It took us the rest of that night and the next day before we were able to relax and enjoy life on board again while cruising to gain information for our wonderful country. We were all rather tired and still slightly nervous so we remained careful. We were a solid crew proving that we could, and would, and did, get back to that relaxed and confident operating atmosphere.

I was very proud to be part of this team of men, the crew of the USS Stickleback. And I was particularly proud of our Captain, who by his calm, sure, actions during those three hours the previous night, breathed his own confidence into every one of us. We were a safe crew, and we were a blessed crew, and we were thankful to have him in control of our safety.

HOT-BUNKING

Submarine sailors are in some ways even closer than birth brothers. To explain what I mean I'll ask you this: "Would you be okay with someone else sleeping in your bunk all the time?" The old-time diesel boat sub sailors are. Because of limited space on these boats, every square inch is configured to have the bare minimum of wasted space. One of the ways to do this is called *hot-bunking*.

Sleeping quarters for the crew are in the forward torpedo room, the after-torpedo room, and the after-battery room, which is next to where they eat and where the galley is. The forward battery is where the officers sleep, which is next to the forward torpedo room.

Because watch duty shifts are on a rotating schedule, there are always men who are awake leaving empty bunks. The men who are not on duty at those times will share those bunks, called *hot-bunking.*

Space for bunks is so limited in the torpedo rooms that men in the forward and after torpedo rooms almost always sleep either under or on top, of a torpedo. There have been times when men have even slept in the torpedo tubes, although that was never a permanent arrangement.

Nevertheless, it takes a particular kind of closeness for men to share their bunks with others, never knowing when they will or will not be sleeping there.

We also used the bunks to keep our clothes ironed. (Well, sort of.) Our uniforms were generally kept folded, under the mattress, so when we took them out they weren't badly wrinkled. They would have been, had we rolled them and stuffed them into our tiny lockers. After all, where would we have room for an iron and an ironing board?

These are just a few of the things that set sub sailors apart from others. As for the bunks, it's said that the bunk *never gets cold;* hence, the name *hot-bunking.*

THE C.O.

I've been on board three different boats and have served under three different Captains. I loved and trusted each one of them with nothing but my deepest respect. We all had a job to do and it was often heavy to carry, but no matter how long I live, I'll never be able to clearly and fully imagine the load the Commanding Officers of these boats have to carry.

Not only do they carry out orders from the Fleet Command but they have the huge responsibility of taking care of 80 or 90 men, week after week, 24 hours a day, minute by minute, just like a mother who loves and worries over her children.

Yeah, the C.O. has to be a mother hen and top dog all at the same time. He has to do a *better-than-average* job of it too because he also counts on those very same men to get him back home again. God bless every one of them.

NORTHERN BEAUTY

We continued on our northern run, searching and recording everything that happened. But some of the things we witnessed will never be recorded in military records. That's the other side of being out in the ocean, seeing things most people never get to see because they never have the chance to go there.

I'm talking about how amazing it is to stand watch on the bridge at night amidst a dark ocean and sky, while clearly and vividly seeing the Aurora

Borealis, (also known as the Northern Lights). With no buildings, no city lights to interfere, and nothing but a clear horizon as their palette, the lights dance and wave as if in a beautiful, colored ballet just for us to watch. The beauty God has created is so marvelous that it outdoes anything man could ever try to make happen.

Another beautiful sight seen on some evenings is just after the sunset when the moon is beaming down enough light by which to see: on those nights the stage is set for a fantastic *dance of the flying fish*.

These *flying fish* leap out of the water and sail across our forward deck, sometimes actually landing on it. There may be as many as fifteen or twenty fish that come out of the water at one time, sailing across the boat in a kind of stage performance of marvelous dancers. Then all of a sudden, they are gone.

Flying fish can't fly the way birds do. But they do make powerful leaps out of the water propelled by their long, wing-like fins, causing them to travel great distances above the water's surface. The reason they do this is to escape from being eaten by other, larger, creatures living in their midst and planning to make a meal of them.

For me, it always broke the still nights on the bridge but it gave me something else that few people have; it gave me stories I could tell my friends when I got back home: the symphony of the flying fish.

NOT ALWAYS SMOOTH SAILING

Things were not always as calm and peaceful as that, however. I still remember the many hours we spent out in the middle of the ocean. At times it seemed like we were confronted by Poseidon, that mythological *King of the Seas* who didn't want us sailing out there. On those nights, the serenity of the Northern Lights or the ballet of the flying fish was gone. It was replaced with an angry ocean filled with power that not even our *iron ship* could control.

On those occasions, and especially at night, when the sea turned its intense, fluid power and fury loose, there was no pleasure in watching it rise up. If we were on lookout and the wind became strong, and the waves even stronger, it would send a chill down my spine. As powerful waves picked up the bow of the boat like some mystical hand was in charge, and then slammed us back down again, it was very normal to fear what it could do to us.

If it became too rough we would go below the raging surface and ride it out underwater. I must be honest with you. While standing bridge watch, if the seas became horribly rough, fear would well up inside of me. The awesome force of the wind and the waves created a fear in me that has stayed with me through the years and has never left.

Since then, I have never taken for granted the omnipotence of God, the creator of it all. Water is more powerful than most people give it credit for. Ask anyone who has weathered storms like that. I'm sure they'll tell you those memories will stay with them forever.

ANOTHER STORM

Many years later on the night of November 6, 1977, while I was at Bible college at Toccoa Falls, Georgia, the dam above the college broke sending millions of gallons of water pouring down the 186-foot-high waterfall, roaring through the campus at 150 miles per hour. If there was ever any doubt in anyone's mind concerning the power of water, all they'd need to see would be the aftermath of the force of that water. It was total destruction.

It was a force that man can only *imagine,* and even then, not fully. The campus was partially destroyed and the lives of 39 innocent people were taken. My daughter was one of them. How the rest of us survived can only be explained by God, Himself.

Water *is* a mighty force and I saw it for the first time at sea while riding out a very small part of a storm, those many years before.

CHAPTER EIGHT

MISSION COMPLETED

Over the next several weeks, photographs and sonar information relating to dozens of Soviet ships were collected before we headed back to Japan and onward toward home. It was a relief for the tension to be over. I guess the trip was pretty much the same as the other northern trips but we were glad to be done and making our way back to Yokosuka and then on to Hawaii.

After many weeks at sea with much stress and anxiety, we finally ended the patrol and were coming into the docks at our home port in Pearl Harbor. It was mid-October, 1956. We pulled into the docks and the men on board were manning the lines fore and aft and were throwing the *heaving lines* to the men standing on the dock who were waiting to tie us to the cleats.

The *heaving line* is a small nylon rope, about the thickness of a clothesline with a lead ball bound and tied with rope at the end of it. This is thrown to the dock and used to pull the larger mooring line attached to it. Then the mooring line is tied to the cleats on the dock and secured to hold the boat in place.

WHOOPS

I was one of the guys to throw the *heaving line* ashore and Dave, my Division Officer, was standing pretty close to me. He was close enough so that when I did throw the line, I somehow knocked his hat off and it went into the water and sank out of sight.

Once again I thought I was in for it in a big, bad way. After all, the hat for an officer at that time probably cost around fifty to a hundred bucks. I knew I was in deep trouble no matter how friendly Dave and I had become.

I looked at him and he looked at me and then he busted out laughing. I was shocked! There was nothing more, ever said about his hat. It was never talked about and I was never asked to pay for it. Some friendships formed on the boats are closer than between birth brothers regardless of rank.

Even with Dave's hat at the bottom of the harbor there at dockside, it was really good to be back home where we could breathe easy. Being home and going out on *daily operations* gave us a renewed and slightly more relaxed time to enjoy the *family feeling* of being on board a boat. Boat sailors or *bubble-heads* as we are sometimes called, are the best of the *best guys in the world.* And I still have contact with some of them.

LOADING DAY

Years ago I heard that the great and well-known leader, Napoleon Bonaparte was credited for saying, "An army marches on its belly", meaning that an army had to be fed. Whether or not that is true I don't know, but I do know that before making a cold war patrol every *department head* was required to check all necessary supplies for however long the duration of the cruise was expected to last.

This included enough food, medical supplies, tools, clothing, and other expendable items we carried to last for the estimated time we would be gone. So it wasn't just food stock we needed but also the armament such as torpedoes.

Additionally, all systems were checked thoroughly to be in perfect working order. If necessary, torpedoes were loaded right along with our full supply of food.

The day of loading torpedoes was a big deal and plenty of *brawn and brains* were needed to do the job efficiently and safely.

Due to the size of a torpedo and the limited space, and depending on how many torpedoes were needed, the job generally took the better part of a day. Plenty of manpower and rigging equipment was required and safety and care were of prime importance.

The chore was generally taken on by the deck crew and torpedo men who knew how to rig and assist in the job. A Mark 14 torpedo, 26 feet long and weighing over one and a half tons was not something to be careless with. Damage was to be minimized or avoided and more importantly, a wrong move was to be prevented. If a torpedo was mishandled it could turn into disaster; it could explode. That was always a possibility and every man was aware of it.

Photo Credit: John Hummel, USN
Loading a Torpedo
http://navsource.org/archives/08/08481.htm

Then, once we were loaded and ready to put out to sea, we always had the disputed privilege of sleeping either on or under a torpedo. Many people (civilians) not familiar with the submarine lifestyle, or uninterested in the pride of serving on a sub would consider this an inconvenience. But for those of us who viewed ourselves as privileged, we found it to be no inconvenience at all. It was our *home* and we loved it.

Near the end of August, we once again had our squeaky shaft inspected. The Stickleback was then ready to be steamed out of Yokosuka and we were on our way back north for our fourth cold war patrol.

Photo Credit: Royalty Free - CorbisImages.com
Hot-Bunking
Posted by Pinterest member: Roberta Caywood

SHOOTING WATER SLUGS

Submariners typically check out the torpedo tubes once each week to make sure they're operational. They are filled with seawater by opening the outer doors, allowing them to fill up, and then they're fired as if there was a torpedo in the tube. This is called *shooting water slugs.*

On older submarines like the Guppy Snorkel, Fast Attack, Killer boats, and all the rest that came before the Nuclear boats, the tubes were fired with 3000-pound high-pressure air. It's pretty loud inside the boat when that happens. The sound can be heard all through the vessel by everyone. So before firing the tubes an announcement is made over the main intercom, "Shooting water slugs."

It's pretty serious business and everyone is on their toes at that time. But, it doesn't mean there can't be a bit of fun while the shooting occurs . . . especially when there are new men on board.

FRESH YOUNG SEAMEN

A favorite submarine prank while at sea is to explain to a *fresh-from-sub-school* seaman how *little creatures* (slugs) get into the bilges from pump drains beneath the engines and how they grow there and have to periodically be removed.

Because we can't surface and go topside just for that little duty, the only way to get them off the boat is to shoot them out of the torpedo tubes. Hence, *shooting water slugs.*

The seaman is handed a bucket and told to go to either of the two engine rooms and get a bucket of *water slugs* from the *bilge,* a small hard-to-reach place beneath the engines. Of course, the machinist on duty is always in on the prank.

The young sailor then heads to the engine room, bucket and scoop in hand, and with permission, he will scoop water from the bilge into the bucket. After a couple of tries, he tells the machinist on duty that he can't see any slugs in the bucket. The machinist usually tells him three or four times to, "scoop a little deeper". But he still doesn't get any slugs.

The point? There *were* no water slugs in the bilge. After having laughs all around, the young, seaman is now qualified to be told what *shooting slugs* was all about.

It was an innocent prank and everyone had a good laugh with these new shipmates who were a little embarrassed, but also a bit wiser. Next time, it might be him who will send someone back to the bilges to get a bucket of *water slugs.* Good, clean, safe fun is a way to keep morale high and keep boredom to a minimum.

Being on the boats at sea day after day in strange places, never knowing what may come next makes it crucial to have some form of levity to keep everyone at peace and keep their minds on the importance of their mission. Water slugs help.

BLUE SUEDE SHOES

Another way we'd break the monotony was to occasionally have a talent show. At times it was quite good but then with folks doing whatever they thought was talent or was entertaining to the rest of the crew, it could be pretty boring.

While I was on the USS Capitaine, SS 336, Ty Hurt, an engineer, Jimmy Dean, a radioman, and I, would often get together in the radio shack and listen to the most popular songs of the day. We'd do our best to sing along and it broke the tedium; plus it brought us together for a little fun.

Ty was the serious type, though, but Jimmy was a cutup. I'm not sure we were all that entertaining but eventually, we were asked to sing at the talent contest. So, we gave it a shot.

The most popular songs we could get at that time were *Blue Suede Shoes* or *Heartbreak Hotel,* both by Elvis Presley, and *A White Sport Coat* by Marty

Robbins. I guess these were some of our favorites. Even today, Marty Robins is still one of my favorite singers.

The three of us sang together as a trio but we sang separately as well, and we had a good time at it. Later on a black guy and a white guy did a duet and they called themselves *Night & Day.* They were very good together and everyone got a kick out of the name they chose.

I have many great memories that were built through my years on both of those boats. I'm reminded of those wonderful times when we all depended on each other for our safety and our very existence. I'd love to live those times all over again.

THE INTERNET

Many years later I found a way to connect with these old friends. One man found my email address, then wrote and asked questions to verify who I was. It worked.

It was our old head cook and Commissary Chief, "Ma" Phelps. He served on the Stickleback and he found me through my email address that he received from other *bubble-heads.*

Once "Ma" made sure I was the right guy, he shared two other addresses with me; the two guys I sang with in the trio. Ty Hurt was in Dallas, Jimmy Dean was in Oregon, and I am in Georgia. We began emailing each other.

Time may have passed but our friendships have held tight and true down through the years. That's what the submarine family is all about. We knew and accepted the fact that if necessary, we would give our lives for one another. The years and the miles apart have not changed that. We share a bond that will never be broken.

"MA" PHELPS

Don't let it be said that "Ma" Phelps wasn't ready and willing to pull pranks on folks. He had a sense of humor for gags which he probably practiced for many years on several new men coming aboard the various boats he served on.

I don't know how he did it but "Ma" always kept a can or two of raw oysters around. I don't think they were used for food very often, but they sure were used for pulling practical jokes.

When we were at sea and the weather would get rough and bumpy, "Ma" would slip someone a couple of raw oysters. That person knew what "Ma" wanted them to do. They would take them from "Ma", unnoticed, and pretend to have sneezed into his hand. Then he would shake what was in his hand onto the table as if it came from his nose. *Gross!* Sometimes it was enough to make a new man sick to his stomach.

Another example was when the boat was rocking and rolling with the waves, someone would get food in his mouth, chew it all together, and begin acting as though he was going to vomit. You need to know this: men who have not been to sea before, have a strong tendency to get seasick. As the boat rolled from side to side and up and down, getting worse by the minute, the more prone they were to vomit their last foods.

"Ma" would nod to someone, giving them his permission to take a couple of spoons full of food into his mouth--without the new man noticing--chew it all up and begin making guttural noises, pretending to get sick. When he saw the new man getting queasy, he would make vomiting sounds and spew it out onto the mess table, right where the new man was sitting. That was all it took for the him to lose his stomach for real.

"Ma" loved pranks but this was allowed only if he gave permission and knew that everyone was *in* on it because he didn't want everyone at the whole mess table getting sick. That's because "Ma" would be the one responsible for cleaning it all up.

Not so funny, you say? Well, maybe not. I suppose you would have to have been there to experience and appreciate it. When the new man had a few months at sea himself, it became funny to him, and soon you'd see him pull the same stunt on the newest men on board. It was all good-natured fun and it kept us from losing our senses when things got monotonous, which wasn't all that often.

CHAPTER NINE

NOT THE WAY IT STARTED

Another bond of friendship that has lasted over the years is my friendship with Mac Ordonio, from Manila in the Philippines. Mac was a Steward's Mate which meant that he was the guy who took care of the officers' quarters, their clothing, their meals, and their general welfare.

From the very first time we met, Mac and I did *not hit it off* so to speak, which is unusual for crew members on the boats. It was apparent to both of us that we rubbed each other the wrong way. I don't know why but for some reason things went from bad to worse.

Mac was about five feet two or three inches tall, with brown Filipino skin, dark hair, and a thick Filipino accent. I have no idea why we didn't get along but we eventually came to the place where late one afternoon we were in port and got into a serious confrontation up on the dock.

One tough word led to another and then we were in a fistfight. I was no slouch when it came to fighting. I had been in numerous street fights before joining the Navy. But I was no match for this short, brown whirlwind called Mac Ordonio.

Three times he knocked me down and twice I got back up. The third time, as I lay on the ground looking up at him he said with that strange, broken, hard-to-understand accent, "Hab you had enough?"

I looked up at him and said, "No . . . Maybe not tonight, not tomorrow, and not the next day but we are going to do this until I have you on the ground looking up at me. And when I do I'm going to put you in the hospital because I'm going to stomp on your face and kick your ribs in."

"FRIENDOS"

Mac stood there and stared. Suddenly, while he was speaking, he put his hand down to help me up and said, "No . . . you will neber do dat to me."

I didn't quite understand why he was offering me his hand. I thought it was some sort of trick, so I waited.

Still holding his right hand out, he said, "You can trust me. Take my hand, will you?" and he helped me get up off the ground.

As I was getting up I said to him, "You are so stupid! Because next time it will be me looking down at you, and you *will* be needing to go to the hospital."

Mac said, "Like I told you, no. You will neber do dat to me because I look into your eyes and I belieb you . . ."

Then, he held his hand toward me with the little finger extended and said, "Do dis."

I wasn't sure what to do so he showed me how to hook our little fingers together. As I slowly reached out to him, he said, "Dis means we will be all-la-la-time friendos. No matter where we eber are, or when, we will always be friendos. Nothing will eber change dat."

Somewhat reluctantly, I hooked pinkies with him, and in that crazy, intimate exchange, we bonded. A friendship was born that would last for a life time.

Mac and I served in peace and friendship together on the Stickleback for several years after that and never had one sarcastic word for each other. We truly did become extremely good friends.

DOES SIZE MATTER?

Mac was little, but he was more than just loud; he was *bad to the bone.*

We had gone on liberty together one night at one of the other islands; Maui, I think it was. As we were heading up a long sidewalk toward a bar to do some more drinking, three Marines were walking toward us, taking up the entire width of the sidewalk. We had already had a few drinks elsewhere but we were sober, in a good mood, and not expecting what came next.

The three Marines divided, one on each side of us and one right down the middle. As they brushed past us, each of them punched us with their elbows, laughing and making ugly comments about sailors.

Mac and I responded by turning around, facing them, and firing a choice comment back about Marines. When we saw it was going to be either fight or flee there was no choice left to be made. We had to fight.

Mac very calmly turned to me and said, "Do not worry. I hab dis cobered. Dis is mine . . ."

While he was speaking he was also removing his black web, Navy issue belt, wrapping a turn of the blunt end around his fist with the flat polished brass buckle hanging free.

I said, "Oh no, Mac. There are three of them. We do this together!"

He said, "You jes stand and let me do dis. I *can* do dis an I *will* do dis. Jes' watch. We are all la la time friendos, remember?"

With that being said, Mac then turned to the three Marines, "An what do you say you will be goin to do to us?"

THE FIGHT WAS ON

Yes, the fight was on. They stepped up to him and the first one swung a fist and missed. Then Mac started swinging his belt like a cowboy would swing his lasso. In mere seconds, each of them was trying to protect themself from the swinging buckle that was smashing their faces, heads, upper body, and arms.

Seriously, he was like a one-man whirlwind. The fight was over before it even got a good start and the three young Marines hurried away, wiping blood from their faces.

At that point, I thought, "And I was dumb enough to fight this guy three times? And even claim that one day I would beat him? Standing there watching the Marines get away I realized that the size of the dog in the fight doesn't matter. It's the size of the fight in the dog that counts. And Mac was a *top dog.* Yeah, I was so glad we were going to be *all la la time friendos.*

A KNOCK AT MY DOOR

Fast forward six or seven years. I was married and living in Spring Valley, a suburb of San Diego, California. I heard a knock on my front door and when I opened it, there stood Mac Ordonio. You could have knocked me over with a feather.

After his discharge, he had left the U.S. and had gone back to living in Manila, in the Philippines. Mac had two weeks' vacation so he found me and came to see me. To say the very least, I was flabbergasted. The friendship was as true as it had been while we were shipmates on the Stickleback six or seven years earlier.

Mac spent a couple of days with us and I was sort of sad to see him leave. I have never seen or heard from him since but it wouldn't be a huge surprise to one day answer a knock at my door and find Mac standing there saying, "You see? I told you we would be all la la time friendos."

Mac, we truly are and always will be the best of friends no matter where we are, no matter how many years pass. I will often think of those times we served together and wish I could see you again. That's what *true friendships* are all about and the submarine service is full of them.

THE TOILET MIGHT EXPLODE?

While having some fun to break up the routine, there are some very serious things to know about submarine life. One of them is that simply flushing a toilet at the wrong time can cause it to explode.

A submarine has to discharge its sanitary tanks out into the sea every so often. On the older boats, they pressurize the sanitary tanks with high-pressure air and blow the contents overboard. It's based on the same principle as firing a slug of water from a torpedo tube.

Therefore it is critical that you *never* flush a toilet while a tank is being blown overboard. Why? Because the toilets are flushed by opening a ball valve that lets the toilet contents flow down into the holding tank, which is then blown out to sea. So if you happen to open the ball valve while the tank is pressurized: BOOM! The toilet can explode from the air pressure coming back through the valve and everything you meant to flush will also come right back . . . right at you!

It's not only unpleasant but quite dangerous as well. That's just one of the many reasons for requiring a probationary period. When a new seaman comes aboard a boat, he is given six months to *qualify.* And there's nothing I can say about qualifying that will make it sound easy. Qualifying is tough, and I'm sure it's supposed to be. But from my personal point of view, the pride and satisfaction I received from it make it all worthwhile.

THE TESTING

To qualify as a submariner, a sailor has to know every system on board: electrical, high-pressure air, water, fuel, navigation, radio, and so on. He has to know how to operate each system safely and properly, so he can step into that job if something should happen to the man who normally fills that position.

For example, have you ever wondered how so many men can continue to breathe once they've been submerged for a long period of time? Sometimes days at a time? Have you wondered what keeps all the oxygen from being used up? The answer to those questions must be known by every freshly qualified man.

He must know how critical it is to have a good oxygen supply. There are no exceptions. He needs to know that submarines make their own oxygen, and how that is accomplished. That's right. They make their own oxygen.

On board any submarine there are always one or two oxygen generators that take in the salty sea water and break it down into molecules and then break those molecules down into their oxygen and hydrogen atoms. Then the oxygen is stored in oxygen banks for emergency use, in case both oxygen generators go down.

If the submarine sinks and loses power, it can bleed oxygen from the storage banks back into the air, using it to sustain life until they get back to the surface. We never want to face that as a necessity, but if it becomes so, the life-saving oxygen is there to sustain every man on board.

THEY CAN DO IT IN THEIR SLEEP

A man fresh from sub-school is allowed six months to be taught and trained, and become intensely familiar with every system on the boat, why it's there, and how to use it. He is taught by other qualified men who, as part of their duty and their job, teach him to be able to step in and take over someone else's job if the need should arise.

It *must* become so much a part of him that even if an emergency arises while he is asleep, he will immediately come awake to perform his emergency duties without making a mistake. Every life on board depends on him, and he depends on them. It *must* become as much a part of him as his breath or his heartbeat.

LET'S HAVE SOME FUN

Once when I was home on leave at Mom and Dad's house, I fell asleep on the couch. My younger brothers and sisters knew just enough about what went on in a submarine to decide they would have some fun with me. (Terrible idea.)

While I was asleep they got down close to my ear and shouted, *"Dive! Dive!"* Needless to say, I came up out of my sleep and began grabbing at the wheel (which wasn't there) to close the bulkhead flapper (valve) and I was twisting it closed as fast as I could, just as I was supposed to. The problem was that I was not on board my submarine. I was sleeping on the couch.

In the mode that we were trained, even while asleep, we had to rise and hurry to get the valve closed; that was essential. Thinking I was onboard, I tore the Venetian blinds to pieces, trying to twist the wheel (which wasn't there) to close the valve.

Funny? Yeah, it was to them, but not to me. I did scold them but realized what happened was all in fun. After that, Mom scolded them, too. An understanding was imparted to my brothers and sisters as they learned that's the way a submariner is trained to operate.

We all knew that was the reaction they'd hoped for and they got exactly what they wanted. Awake or asleep, every man on board *must* be alert enough to react properly and immediately to all emergencies. Many lives depend on it.

IT'S NOW OR NEVER

At the end of the six-month learning period, the man being examined is walked through the boat, having question after question shot at him by the testing team. This team usually consists of three or four men who will actually--at various points--blindfold the man being tested. Some of the questions are shot at him as if it was an emergency situation. The questions cover *everything* that goes into making up the boat and its operations. *It is a must* that he passes the test or he is removed from submarine service and put either on a surface ship or on a shore base.

THE DOLPHINS

Qualifying is rather hard but when he is fully qualified and finally receives his *Dolphins* which are pinned on him, the pride runs deep and the pride never dies. Such is true for every man going through this process.

He knows he has worked hard and passed all the tests to become part of a very elite group of military persons. And with that pride and satisfaction, there is immediately birthed within him a strong sense of responsibility.

Photo Credit: Silver Submarine Dolphins is a mixed media by Baltzgar https://fineartamerica.com/featured/silver-submarine-dolphins-baltzgar.html

Once he qualifies, then, and only then, is he allowed to wear the *Dolphins* on his chest, telling all the world that he is smart, he has persevered, he is dedicated, and he is privileged to serve on board a United States Navy submarine. He has become a *man of the boats.* Dolphins are an honor to wear.

Today, on those special days of remembrance to our service men and women, such as Veterans Day, Memorial Day, and other national days set aside for our military personnel and their families, I still put on my uniform and wear my Dolphins with a special sort of dignity.

I once heard a phrase describing the Dolphins this way: "These Dolphins--once you pin them on your chest--leave deep marks right over your heart, long after the uniform has been put away."

Now that I am many years removed from being on active duty, and my uniform is put away, except for special occasions, I must absolutely and fully agree with that well-spoken sentiment. I couldn't have said it any better. Submarines run deep, but so does the pride of the men who ride them to protect our great country. And the Dolphins on their chest attest to that.

CHAPTER TEN

WE ALL HAVE TO LEARN

The Junior Officers have their problems too. They are all new at some point and have to learn their jobs just as we did. One of those jobs was to remove the boat from being tied to the dock and then navigate it through the harbor and out to sea without incident.

The same applies when coming back into the dock. Everything must work exactly right to prevent mishaps. But sometimes things can go wrong.

A Junior Officer on another boat was in charge of bringing it in from *daily operations,* but he was bringing her in too fast. The bank in front of the dock is a sloping, earthen bank, covered or protected by wood. It is meant to keep a boat from getting too close.

Because the officer was coming in too fast, he hit the bank and it sent the bow of the boat up the incline, into the air, causing it to hit a car parked on the near edge of the dock. The boat then slid back down and neatly parked itself at the dock where it rested. It's amusing now but at the time, there was nothing funny about it. There was no damage to the boat but it sure did cause damage to the automobile.

The insurance covered everything but the word went around that there was going to be a car painted on the side of the *sail,* just like airplanes have on the side of their planes when they shoot down an enemy plane.

So if you ever see a picture of a submarine with a bright red car painted on the side of her *sail*, up there where the numbers are, you'll know why. No, the painting never happened. Just having the story going around was embarrassment enough for this Junior Officer. He was razzed for a week or two but it was all done in good fun and then it was forgotten. I'm sure he eventually made a good officer.

A BOAT IN "MOTH-BALLS"

The years I served on the Stickleback are years filled with many great memories. But life goes on and it was time for me to serve my other more personal needs.

It was early summer 1957 and one of those needs was that I wanted to go back stateside so I could go home and get married in August.

Almost by fate, I learned about another submarine, the USS Capitaine, SS 336, that was being taken out of *mothballs* at Mare Island, in Vallejo, California, and put back into service.

As mentioned earlier, *mothballs* is a term used to describe a boat that's put into *reserve.* It's no longer in active service but it's not destroyed. It is available to be brought back again if the need arises. Since I was going to start my married life--a home life away from the Navy--I thought transferring to a boat in her home port in California would be a good way for me to stay in the U.S. It would keep me from going on any more war patrols far away from home.

I volunteered for that duty and easily got my transfer orders and left for Vallejo. I would be leaving behind not only my wonderful duties, stationed in Hawaii but also many amazing memories of serving aboard the Stickleback and being with my terrific shipmates.

I felt if my reasons were neither selfish nor harmful, then I was sure my life aboard the Capitaine would continue taking me where I was supposed to be, just as well as before. I would find out soon enough as I sat aboard my military flight back to the States with Mare Island as my destination.

HOLLYWOOD

The background on the Capitaine is somewhat interesting. Here is just a bit of it which you might find enjoyable.

In the past, all Submarines were named after some form of sea life. Today, that isn't always true; I wish it still was. Now, they're sometimes named after politicians and other supposed famous people.

The Capitaine is a brilliantly colored fish found in the waters of the Atlantic Ocean from North Carolina to Panama. Why the boat was named after that particular fish I have no idea, but that's where the name came from and I'm perfectly fine with that.

The Capitaine was built by The Electric Boat Company in Groton, Connecticut, and launched on October 1, 1944. After operating in shakedown cruises for a couple of months, the Capitaine left the East coast and got underway from New London on March 7, 1945. She arrived at Pearl Harbor on April 15 and entered World War Two.

The Capitaine hit the ports of Subic Bay, Borneo, Australia, and the Philippines and was decommissioned and placed in reserve at Mare Island Naval Shipyard on February 10, 1950.

In 1951, Hollywood wanted to make a movie called *Submarine Command*, and they were granted permission to shoot their movie on board the Capitaine. The actors starring in it were William Holden, William Bendix, Jack Kelly, and other well-known actors.

In that movie, she portrayed the USS Trigger, SS 237, but someone in Hollywood didn't do their job very well in preparing her. They had forgotten to change the hull number, so in the movie, the number 336 could still be seen on the *sail.*

I guess Hollywood liked the results of the movie because in 1964 they requested that she be featured in another film called, *In Harm's Way* starring John Wayne, Kirk Douglas, Dana Andrews, Slim Pickens, and many other famous players. I've seen both of those movies and I quite enjoyed them.

If you ever have the opportunity to see them, I hope you will because they show a great deal about the type of men who go to sea in these submarines. They are all men of heart and courage beyond what even they know they possess.

Somehow sub-service brings out the very best in a man who falls in love with and serves aboard these amazing and marvelous U.S. Navy submersible boats. There's nothing else quite like them to a man who wants to serve his country with a unique pride in a most adventurous way. *Bubble heads* are what they call us. And that's just fine with me.

Photo Credit: USS Capataine (SS-336) Courtesy Subnet.com
http://www.navsource.org/archives/08/08336.htm

MY NEW HOME

With the approval for my request to transfer from the Stickleback to the Capitaine, I was flown from Pearl Harbor to Mare Island, at Vallejo. From there I would be able to get a 30-day leave from the Capitaine and go home for my wedding. Until then, I became part of the crew who were going to pull the USS

Capitaine, SS 336 out of mothballs and recommission her.

We did everything from scraping off old paint and applying the new to checking, rechecking, double-checking, and even triple-checking every electrical and mechanical system on the boat.

Then it was checked again. After all, when she was put to sea the lives of more than 80 people would be depending on her to be fully functional and safe in every respect.

After several months of going over every system to the best of our ability and being sure each one was working properly, we considered everything ready to be tested for sea duty.

There were many more tests needed with her actually in and under the water. That couldn't be done sitting in the Navy yard, so she had to launch.

We finally came to the place where everyone believed the Capitaine was ready for a month of sea trials. We Wanted to check out the work that was done so, she put out to sea.

For the first leg of trials, Captain, LT Eugene Dente, our Commanding Officer, and others who had authority and concerns weighed in and said they were satisfied. We then reported to the Pacific Fleet, in San Diego which was to become her home port.

Captain Dente was nicknamed *Captain Hook* but I never did find out why. Did it fit him? I don't know. But we never did say it in front of him. Personally, I liked him. He was a good C.O.

TENDERLY TIED

While we were in San Diego, we took part in training exercises, served other ships as a target, (pretend of course) served in anti-submarine exercises, (safe exercises like sending flares), and made a trip to the Far East for Seventh Fleet duty.

Also from home port San Diego, we trained Navy Reservists who thought they might someday be interested in Submarine duty. While at port at San Diego, we were generally tied up next to the USS Nereus, AS 17, a submarine tender, meaning she was a support ship for submarines, both in port and at sea.

When a submarine tender is in port she is anchored in the harbor, and submarines will tie up to her, (as seen in the photo below), and men can go from there into the city by a small boat called *a shuttle boat.* This prevents the submarine from having to get to a dock. Many ports don't have any place made available for them, anyway. (Is that because of the cars at dockside?) Just a joke.

Photo Credit: U.S. Naval Institute (Archives)
USS Neureus with Submarines in Tow
https://photos.usni.org/content/1054286jpg

Photo Credit: Donald M. McPherson, Corte Madera, California, 1969; USS NEREUS (AS-17) https://www.history.navy.mil/our-collections/photography/numerical-list-of-images/nhhc-series/nh-series/NH-69000/NH-69846.html

CAROLINE AND EDDIE

One afternoon I was on the Nereus and I bumped into a guy. Literally! I guess neither one of us was paying attention so we apologized and turned to go our separate ways. He stopped and turned around about the same time I did. We stood there for a minute just looking at each other when finally he said, "Where do I know you from?

At the same time I was saying to him, "Shouldn't I know you?" I thought he looked familiar so I began naming different places I had been and the different boats I had been on.

I explained, "First there was Commissary school and the 35 Millimeter Camera Operator school, both right here in San Diego. Then, I went to Submarine School in New London. Next, I was on the K-2 and the Stickleback, both in Hawaii. Then, I came back to the states to Mare Island at Vallejo, and now I'm on the

Capitaine. We're tied up right next to you. That ring any bells?"

None of it rang a bell but suddenly, he brightened up and said, "Hey. Where are you from? I mean where did you go to school? Where did you grow up?"

Then everything clicked for both of us. I said, "Eddie!? You're Eddie Feiler. Right?"

Yeah, it was him alright. We had gone to high school together but his family had moved away to Arizona, I think. I told him that pretty soon I was going back home on leave to get married. With that, he dropped a bombshell on me.

FINDING A HOME

"Yeah? Well, I'm already married! And you probably know who she is. I married Caroline Lutz. You have to come home and meet her. We're living right here in San Diego."

I was surprised because I did know Caroline when she and I had been in several classes together in school.

Then I told Eddie what I was going to have to do about getting my wife here after we were married and how I would be leaving for Vallejo right away so I had to get a place to live for her before I left.

He said, "Oh no. Don't be in such a hurry. Wait until she is here so she can help you find a place to live that both of you will be happy with. Just bring her to Caroline and me. Leave her with us until you get back. We'll take good care of her for you."

So that's the way it was going to take place. I would go home, get married, come back to San Diego, leave her with Eddie and Caroline and she would drive me to Vallejo where I would board the Capitaine and go to Port Angeles. Then she would drive back to San Diego and we would get together there later and find a place to live.

It was enough to make my head spin but it sure solved some problems about where my wife would be while I was gone. I could hardly wait for the roller coaster to stop.

THE MERRY-GO-ROUND

From San Diego, I was given *leave* so I went home for my wedding. While I was home, the boat left San Diego and went to Mare Island and had a few system glitches checked along the way (as the first part of her real *shake-down* cruise).

When I got back from my *leave* I took my new bride to San Diego so she could meet Caroline and Eddie. Then I continued on my way to board the Capitaine in time to make the trip from Mare Island to Port Angeles, Washington, home of the largest Coast Guard station in the country. My plans were working out just fine so far.

My wife drove me to Vallejo where I boarded my boat. Then she drove back to San Diego, alone, and stayed with my friends until I got back. Together, we would then find a place to live. Simple? Yes, but hindsight now tells me it was all a big mistake.

I say this because even though we found a place to live and life went on its *married* way, there was no peace in it.

CHAPTER ELEVEN

SHIP FOR SIX MORE

When my enlistment was almost up, the Captain called me into his stateroom for a *sit down* about my future with the Navy. He wanted me to stay in submarine service and to advance in it. He suggested that since I already had four years in, I had a good start on a 20-year retirement career. He told me the Navy needed a good torpedo man and with that in mind, he had an offer that might interest me.

He offered me two years of college, two years of nuclear fission school in Orlando, Florida, and two years of active duty if I would *ship for six;* (that's re-enlisting for six more years).

After completing college and nuclear school, I would go into nuclear submarines for the final two years of my six-year re-enlistment. I was excited to say "yes", but I told him I would have to talk it over with my wife.

The opportunities he offered me were great. At the end of that six-year re-enlistment, I would have ten years in, which would be halfway to a twenty-year retirement. It was a terrific offer and I appreciated his confidence in me. I told him I appreciated *him,* and I think that was the only time he ever shook hands with me.

I could have a good life in the service and I was glad to be given this chance to excel. After a twenty-year career, I would have a lifetime retirement income. What a great financial support system that would be for my family.

There was just one problem. My fresh young wife threw a fit.

THE CHOICE WAS MINE

After discussing and cussing the offer, the many arguments we had were her way of taking control of my life.

She said, "It's either the Navy or me! Either you get out and we go home or I go home alone and you can do whatever you want to. Period."

Well, ignorance *is* as ignorance *does.* I gave up the twenty-year Military career I wanted, and a few months later when I was discharged, we went back home. It didn't take very long for me to realize, *that* decision was the wrong one.

Two years later the marriage was *rocky and troubled* anyway. In an attempt to save it, we decided we would leave the influence of family and find a place to start over. I was young and I was dedicated and I did want to save my marriage.

WOULD A NEW START WORK?

We left Illinois and moved back to California. Somehow we landed in Downey. Downey is about 15 miles south of Los Angeles and it's the birthplace of the Apollo Space Program. It's also home to the oldest McDonald's in the world. But to me, it was just another pretty little California city where I got a job in construction. I was just trying to make my young marriage work, whatever it took.

Late one evening, while at home there in Downy, my wife and I settled into a very intimate yet odd conversation. It created a very strange atmosphere. I say strange because we were able to bear and share our deepest feelings without screaming, yelling, and fighting.

It almost seemed to be staged, (and probably was). We entered into a very open, heart-to-heart talk about where our marriage was and whether we were going to make it or not. This was unusual for us.

Looking back, I can only guess it was a time for her to unburden herself from her past and give a full confession of her secrets. They say that *confession is good for the soul* but I think at this point she was willing to confess and unload because she just didn't care anymore.

A SAD REVELATION

She began telling me about when she dropped me at Vallejo so I could join back up with the Capitaine and head for Port Angeles. During that once-ever, strange, chosen discussion, she confessed and almost bragged that when she headed back to San Diego, she had had the time of her life.

To her way of thinking, she was footloose and free as a bird and she was going to enjoy herself. The very first night without me, she drove back to San Diego and met a guy at a truck stop and spent the night with him. That was the beginning of a long list of adulteries, too many to remember, which she apparently felt no guilt in telling me about.

Not only did it hurt me deeply but it also made me suffer to realize I had not taken that opportunity to re-enlist and stay in the Navy. I loved submarine service and I wanted to stay, but I didn't.

Now, many years have passed so the past is where it will have to stay. Yet, not staying in the military and accepting those great opportunities is something I'll always regret. It would have been life-changing for me.

MY FRIEND MARVIN

Not all of my memories are sad. There are many good memories from those days too. One of them happened one evening while I was still on the

Capitaine, just before she was ready to make another extended Northern Pacific cruise. Let me tell you about it.

We were tied to the Nereus and I was standing *deck watch* when this innocent-looking, pink-faced young man came up the gangway, saluted, and asked for permission to come aboard. We spoke briefly as he told me he had just graduated from submarine school in New London, Connecticut, and was assigned to the Capitaine. I checked his papers and logged him aboard.

I guess because he looked to me as an older brother, Marvin and I became fast friends and I thought it was just a passing thing because I was about to get discharged and he was just getting started. But I was wrong.

Many years later, (and I do mean *many years later)*, about 50 years had passed when I received an email from someone in Indiana asking how old I was and if I ever served in submarines. Once I saw his name I was pleased to see that it was Marvin.

He had found my address in an email he received from another *submarine buddy* and he emailed the address hoping it was me. We started to email back and forth and he told me how he had thought of me many times over the years and wondered whatever happened to me. That's when he told me two things that have become very special to me.

First, I had been one of the men who took Marv through the boat, running him through his qualification test to see if he could stay on submarines permanently. And he qualified. I didn't remember doing that with him but I'm very pleased that it was important enough for him to remember it.

Second, through the years he wondered if I had ever become a Christian by accepting Jesus Christ as my Lord and Savior. I told him I had and he was very happy to hear it.

Ever since the evening I logged him in to come aboard, he had always looked up to me as a big brother. How nice that was for me to hear him say so.

We have continued emailing each other to this day. It's nice to have that ongoing relationship all these years later.

Marv and I served together until I was discharged. We did lose touch, but Marvin, who lives in northern Indiana, went on to spend twenty years in submarines, and another ten years as an instructor/teacher at Great Lakes Naval Center, the same place I was sent to for my basic training long ago. He did just exactly what I should have done.

THE FATE OF THE CAPITAINE

Update: December 5, 1977, twenty years after I logged Marvin aboard the Capitaine, she was sold to Italy, to strengthen their Naval power. Then she was struck from the U.S. Naval Registry because she belonged to someone else. I hope they always have safe sailing with her because not all submarine stories come out so well.

HOME

On June 6, 1957, just a few months after I had transferred to the Capitaine, the Stickleback left Pearl Harbor on her fifth, and final cold war patrol. During this patrol, she returned to the coast of Siberia and patrolled the shipping lanes between Petropavlovsk and Provideniya, Soviet Union. During this brief patrol, she collected intelligence information on 40 Soviet ships before returning to Pearl Harbor.

Once she was safely back home again, the Stickleback began regular operations, playing war games with our ships in the Pacific Ocean around Pearl

Harbor. She was home and she was safe, and that's always a good feeling.

There's no way I can relay to you the feeling of coming back home. There's nothing like pulling in and tying up to your *home port;* back to where all is peaceful; back to where every day you feel as safe as if you were in your loving mother's arms. *Yes, home.*

Many men have come back home and taken in that wonderful feeling from one's toes to one's gut, hanging there like a special holiday that can't be described. *Home* is where it feels like everyone who greets you, loves you, and has truly missed you while you were away.

This was my last time to come *home* on the Stickleback, and it was also special for her.

HER FINAL PATROL

The Stickleback had made several trips but no one could have guessed the outcome of this fifth patrol into the far northern reaches of the Pacific; it would be *her last trip.*

On May 28, 1958, the Stickleback was participating in an anti-submarine warfare exercise 19 miles off the coast of Oahu, Hawaii where the Pearl Harbor Navy Base is located. She was working with the USS Silverstein, DE-534, a destroyer escort, and the USS Sabalo, SS-302, another submarine, and a torpedo retriever boat. (A torpedo retriever boat is exactly that: a boat approximately 60 feet long that retrieves practice torpedoes used during exercises with other ships.)

The exercises continued into the afternoon of the next day, May 29, when during these drills the Stickleback completed a simulated torpedo run on the Silverstein.

During the run, the Stickleback was diving to a safe depth when she lost power and descended. She went down, uncontrolled, to somewhere between 700 and 800 feet. Negative tank was emptied of all of its water to add buoyancy and the Stickleback was broached (shot to the surface rapidly), approximately 500 feet directly in the path of the Silverstein.

A flare could not be fired from the Stickleback warning the Silverstein because crew members in the after-torpedo room, from where the flare would be fired, had already evacuated the room due to the collision alarm that had been sounded.

Seeing the danger ahead, the Silverstein *backed-down* full power and put her rudder hard left to turn away from the Stickleback. But sadly enough; they weren't able to get it done in time to avoid a collision.

The Silverstein slammed into the Stickleback on her port side, tearing a huge gash in her pressure hull all the way to the internal, water-tight compartments.

People have asked how a strong steel ship could be slashed into that easily. The truth is, the test depth of any particular submarine and the thickness of her steel hull, and many other particulars about submarines are still kept as pretty much secret information. But they can be heavily damaged and even destroyed, as proven by this accident with the USS Silverstein.

LOVE'S LABOR LOST

Once she was hit by the Silverstein, many of the Stickleback's crew were removed by the torpedo retriever boat as quickly as possible. All the other ships in the area also picked up Stickleback crew members from the water.

Besides the torpedo retriever, combined efforts were made by the Destroyer Escort, Silverstein DE-534, the submarine, Sabalo, SS-302, the Destroyer Escort Sturtevant, DE-239, and the Greenlet, ASR-10, an air-sea rescue craft.

They all worked together not only to save the lives of the crew but were also trying to save the Stickleback by putting lines around her in an attempt to basically float and carry her back to port. But it was what may be called *love's labor lost* because all the work was to no avail.

Photo Credit: National Archives
USS SILVERSTEIN (DE-534) and USS STICKLEBACK (SS-415)
https://www.history.navy.mil/our-collections/photography/numerical-list-of-images/nhhc-series/nh-series/USN-1036000/USN-1036229.html

Compartment after compartment flooded and at 1857 hours (6:57 pm) on May 29, 1958, my old friend, the Stickleback, sank in 1,800 fathoms of water, (11,000 feet) making it her last and final dive. She was *deep-sixed* to the bottom of the Pacific Ocean, and that's where she is laid yet today. Thank God, not a single soul has been left on board.

WRECKAGE FOUND

March 10, 2020, 62 years after the accident that sunk her, the Stickleback was discovered by a team of ocean explorers utilizing pioneering robotics and methods at the forefront of today's underwater technology. There, in nearly 11,000 feet of water she was found by veteran ocean explorer and Tiburon Subsea CEO, Tim Taylor and his "Lost 52 Project" team.

The *Lost 52 Project* is a private organization founded by Tim Taylor to research the 52 U.S. Navy submarines lost on patrol during the Second World War. To date, they have found at least 12 of them including the Stickleback.

Their teams are equipped with a combination of autonomous underwater vehicles (AUV), remotely operated vehicles (ROV), and advanced photogrammetry imaging technology.

When they found the Stickleback she was broken into two pieces but there is no doubt it was her. As you can see by this underwater photograph they took, number 415 is just as clear as it can be upon her *sail,* so there is no doubt. It is the Stickleback.

Photo Credit: Lost 52 Project
Stickleback Found
http://www.lost52project.org/Stickleback-Gallery.html

There were many other pictures taken which I declined to share here. I decline to show them because they are just images of what once was a great fighting ship, but now has 62 years of sea growth covering most of her broken body.

The number 415, was not just her hull number. Over the years, it was for so many of us, our home address and we will always miss her.

I must say this once more: "Thank God there are no bodies on board". They were all rescued at the time of her sinking.

So, the Stickleback made her last dive. Not on purpose, but as if she had been *deep-sixed* forever. She is, and always will be missed for years to come by all of us who were happy and proud to call her our home.

CHAPTER TWELVE

PRAISING THE MILITARY

Sir Winston Churchill put it this way: "Of all the branches of men in the forces there is none which shows more devotion and faces grimmer perils than the submariners."

It is amazingly nice to hear such an accolade of praise and appreciation from someone so prestigious as Winston Churchill, who served as a leader during both World War One and World War Two. But I personally, highly admire *every* person who wears a military uniform. Even so, I will always be a submarine sailor.

I can't write about the unflinching loyalty and devotion to duties and responsibilities of our service personnel without also voicing my deep appreciation of a certain *silent backbone of support* for our troops. I'm talking about the wives, the husbands, the children, and the Moms and Dads who were, and are yet today at home, loyally supporting our servicemen and women. That's the full measure of our military strength.

Thanks to every one of you and may God bless you and keep you safe. You are the unseen strength of our people in uniform. We stand behind you one hundred percent.

THANKS TO ALL

President Ronald Reagan believed so wholeheartedly in the importance of military spouses that on May 23, 1984, he enacted what is called Proclamation 5184, which made that date the first Military Spouse Day. That proclamation declares:

"As volunteers, military spouses have provided exemplary service and leadership in educational, community, recreational, religious, social, and cultural endeavors. And as parents and homemakers, they preserve the cornerstone of our Nation's strength—the American family."

Congress would later call it *Military Spouse Appreciation Day.* Each year, the sitting U.S. President commemorates this special day with a ceremonial speech and proclamation of the importance of family support for our troops. The U.S. Secretary of Defense, Caspar Weinberger, standardized the date by declaring the Friday before Mother's Day as Military Spouse Appreciation Day.

To all U.S. service personnel and their families, with humble and honest appreciation, I want to personally take this opportunity to say:

"May God always provide each and every one of you with calm seas and clear sailing as you devote your life and strength to the United States Military, and the protection of our loved ones by your sacrifice. May God deeply, and forever, bless you."

IN APPRECIATION

I once heard it said this way: "Submariners are a special brotherhood. Either all come to the surface or no one does. On a submarine, the phrase, *all for one and one for all* is not just a slogan; it's a reality."

I appreciate whoever said this because having lived it myself, I could not agree more.

So for now, because I've found and traveled the *big blue ball* of my childhood, I'll try to find a good book to read while I lie on my back beneath that shade tree that I enjoyed so long ago, and I'll say to you:

"Thanks for reading from the pages of this book
and thanks from the bottom of my heart."
U.S. Veteran; R. Douglas Veer

Photo Credit: https://www.popularpatch.com/navy-patches/submarine-patches/ssk-2-uss-bass-patch

Photo Credit: U.S. Government/Public Domain https://commons.wikimedia.org/wiki/File:USS_Bass_(SSK-2).jpg

Photo Credit: https://www.popularpatch.com/navy-patches/submarine-patches/ss-415-uss-stickleback-patch

Photo Credit: USS Stickleback SS-415 https://www.reddit.com/r/submarines/comments/8if2l1/uss_stickleback_ss415_august_4_1954/

Photo Credit: https://www.popularpatch.com/navy-patches/submarine-patches/ss-336-uss-capitaine-patch-c-version

Photo Credit: USS Capataine (SS-336) Courtesy Subnet.com http://www.navsource.org/archives/08/08336.htm

EPILOGUE

Being a crew member on board a submarine was an amazing experience, teaching me the value of personal discipline. But life and the pressures of it after I left the Navy took me down pathways I should never have ventured. Wild nights with equally wild people; alcohol, and often drugs became part of that lifestyle.

Though I lived a kind of crazy life for many years, it was never the fault of anyone but myself.

Then, because of the loving, caring heart of a blessed, saintly, elderly lady who lived next door to me in Dallas, I accepted Jesus Christ as my Savior. He forgave me, and He completely changed my life, erasing those many years of offenses.

Since then I have had the peace that passes understanding and a joy so deep it has replaced all my frustrations and doubt. Today, my purpose in life is to see this wonderful life change in others just as I have been transformed.

You can have this same peace and joy if you want it. The following pages explain the gift of life that Jesus offers and the steps you can take to receive it.

Many blessings to you.

R. Douglas Veer

HOW TO BECOME A CHRISTIAN

❧ ❧ ❧

GOD HAS ALWAYS WANTED A PERSONAL RELATIONSHIP WITH YOU...

Even before He created you, He chose to show His love toward you, through His sacrifice on the cross, and through His Word. Here are some very simple steps for understanding salvation and becoming a Christian believer.

ADMIT TO GOD THAT YOU HAVE SINNED

The fact is, everyone sins. We need to understand this before we can go any further.

Roman 3:23

"For all have sinned and come short of the glory of God"

Romans 3:10-12

"As it is written: `There is no one righteous, not even one; there is no one who understands, no one who seeks God. All have turned away, they have together become worthless; there is no one who does good, not even one.'"

Unfortunately, there are consequences for our sins. It is death. We all face physical death but anyone who is not a Christian believer when he/she dies, will die spiritually as well as physically and be eternally separated from God.

Romans 6:23a

"The wages of sin is death . . . "

It's also important to know that we cannot earn salvation based on our own merits. We cannot be good enough, or smart enough,

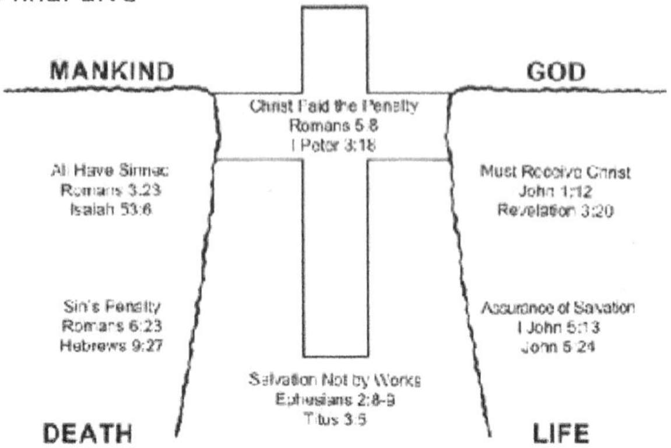

or kind enough, or religious enough.
We are not only sinners, but we fall short
of the standard God requires.
SO, WHAT SHOULD WE DO?

BELIEVE THAT JESUS PAID FOR YOUR SINS ON THE CROSS

When Jesus died on the cross and rose again, He paid the full penalty for your sins and mine, and the sins of every person, living, dead, and even yet unborn.
The only condition for receiving His salvation is that we *believe* He did this for us, according to His Word.
Yes, He did it because He loves us,
and because we couldn't pay for our sins, ourselves.

John 3:16 King James Version (KJV)

"For God so loved the world, that he gave his only begotten Son, that whosoever believeth in him should not perish, but have everlasting life."

Romans 5:8

"God demonstrated His love toward us in that while we were still sinners, Christ died for us."
God's word leads us to believe in Jesus Christ as God's Son and accept Jesus' gift of forgiveness for our sins. Along with God's forgiveness,
we will receive God's gift of eternal life.

Romans 6:23b

" . . . the gift of God is Eternal Life through
Jesus Christ our Lord"

Would you like to receive Jesus Christ into your heart and become a Christian believer today?

Maybe you need some help; a starting point, perhaps? Here's an example of how a person could pray to receive salvation:

THE SINNER'S PRAYER

"Dear God, I know that I have sinned against you and am deserving of punishment. But because you love me so much, you sent your only Son, Jesus Christ, to take the punishment that I deserve. With your help, I place my faith and trust in You as my Savior and Lord. Please forgive me and come into my heart. With your help, I will live for you from this day forward. Thank You for saving me, and for offering me Your wonderful grace. Thank You also for the gift of eternal life! In Jesus' name I pray, Amen!"

In themselves, these words will not save you. But praying in faith and believing on Jesus Christ, will!

Romans 10:13

"Whosoever will call on the name of the Lord
shall be saved."

Romans 5:1

"Therefore, since we have been Justified through Faith,
we have Peace with God through our Lord Jesus Christ."

Romans 8:1

"Therefore, there is now no condemnation
for those who are in Christ Jesus."

John 1:12
"And as many as received Him,
and believed in His name,
to them He gave the power
to become children of God."

Romans 8:38-39
"For I am convinced that neither death nor life,
neither angels nor demons neither the present nor the
future, nor any powers, neither height nor depth,
nor anything else in all creation,
will be able to separate us from the love of God
that is in Christ Jesus our Lord."

Romans 10:9-10
"If you confess with your mouth that Jesus is Lord,
and believe in your heart that
God raised Jesus from the dead,
you shall be saved;
for with the heart man believes unto righteousness,
and with the mouth confession is made unto salvation."

WHAT'S NEXT?
Confess your faith in Jesus Christ
Share your experience with the people in your life
so they may come to know Him too.
Talking about it proves you believe in Him.

* * *

**AS A CHRISTIAN BELIEVER,
REMEMBER TO DO THE FOLLOWING:**
1. Read your Bible daily
2. Pray to the Lord daily
3. Find and attend
a local Bible believing church

Made in the USA
Monee, IL
04 November 2026

67376418R00066